T0215826

BestMasters

Springer awards „BestMasters" to the best master's theses which have been completed at renowned Universities in Germany, Austria, and Switzerland.

The studies received highest marks and were recommended for publication by supervisors. They address current issues from various fields of research in natural sciences, psychology, technology, and economics.

The series addresses practitioners as well as scientists and, in particular, offers guidance for early stage researchers.

Christopher J. Stein

Highly Accurate Spectroscopic Parameters from Ab Initio Calculations

The Interstellar Molecules
l-C_3H^+ and C_4

 Springer Spektrum

Christopher J. Stein
Zurich, Switzerland

BestMasters
ISBN 978-3-658-14829-4   ISBN 978-3-658-14830-0 (eBook)
DOI 10.1007/978-3-658-14830-0

Library of Congress Control Number: 2016945149

Springer Spektrum

Printed on acid-free paper

This Springer Spektrum imprint is published by Springer Nature
The registered company is Springer Fachmedien Wiesbaden GmbH

Preface

As this book contains my masters thesis, it is a summary of the insight I gained in the field of theoretical rovibrational spectroscopy of small molecules. It is more detailed and descriptive than a research article but certainly less educational than lecture notes or a textbook and this book should not be mistaken for any of these types of scientific texts. The way the methods are presented, however, should allow the reader to comprehend in detail the individual steps that were taken to obtain the results presented here. A more detailed understanding can be achieved with the aid of the standard textbooks and original research articles referenced within the text. The theory described in Chapter 2 and the Appendix should allow students of chemistry or physics and scientists that are new to the field to carry out their own calculations on similar systems and might help them to program their own rovibrational perturbation theory program. Astrochemists might be more interested in the results for the two linear interstellar molecules l-C_3H^+ and C_4.

This work was realized only due to the help, knowledge and patience of my supervisor for more than two years: Prof. Dr. Peter Botschwina. Directly after my bachelor studies he offered me to work in his group as a research student and teaching assistant. During this time in Göttingen I had the chance to get a first insight into the scientific community, visit international conferences and get to know the process of scientific publishing. He offered all this generously while demanding nothing but an interest in science itself and his field in particular. There is no doubt that he fuelled my interest in theoretical chemistry and I am and always will be most grateful for his guidance and support. He died, rather unexpectedly, on December 27, 2014. His death is a great loss for the scientific community and the students but both have their invaluable memories.

I am grateful to my current supervisor Prof. Markus Reiher. He offered me a fascinating research topic that I enjoy working on every day and an environment that is extremely motivating. Furthermore, he recommended me for Springer's *Best Masters* series.
I also want to thank Prof. Ricardo Mata, who was the second referee for my thesis and a valuable advisor for many decisions I had to take.
Thanks are also due to my former colleagues Dr. Peter Sebald, Dr. Rainer Oswald, Benjamin Schröder, Arne Bargholz and Oskar Weser.
Studying science would be impossible (at least for me) without friends that find the right balance between motivating me and distracting me with non-scientific topics. In all stages of my life I had the luck to meet exceptionally great people and I am most grateful to all of you!

I deeply thank my family and especially my parents whose unconditional support and love are most precious to me! Without you, nothing I achieved in my life would have been possible.

Preface

Contents

List of Tables

List of Figures

Acronyms

ACPF	average coupled-pair functional
CABS	complementary auxiliary orbital basis
CASSCF	complete active space self-consistent field
CBS	complete basis set
CC	coupled cluster
CCSD	coupled cluster with singles and doubles
CCSD(T)	coupled cluster with singles, doubles and perturbative triples
CFF	cubic force-field
CI	configuration interaction
CSF	configuration state function
CV	core-valence
DBOC	diagonal Born-Oppenheimer correction
DF	density fitting
DKH2	Douglas-Kroll-Hess method at second-order
EDMF	electric dipole moment function
HC	higher-order correlation
HF	Hartree-Fock
IR	infrared
ISM	interstellar medium
MCSCF	multi-configuration self-consistent field
MRCI	multi-reference configuration interaction
MRCISD	multi-reference configuration interaction with singles and doubles
MRPT2	multi-reference perturbation theory at second-order
PEF	potential energy function
PES	potential energy surface

QFF   quartic force-field

RI   resolution of the identity

ROHF   restricted open-shell Hartree-Fock

RHF-RCCSD partially spin-restricted coupled cluster with singles and doubles based on a restricted HF reference

RHF-UCCSD unrestricted coupled cluster with singles and doubles based on a restricted HF reference

SCF   self-consistent field

UHF   unrestricted Hartree-Fock

VPT2   vibrational perturbation theory at second-order

1 Introduction

When trying to identify the molecules whose pure rotational or rovibrational transitions give rise to lines observed in astronomical sources, astrochemists and astrophysicists often face the problem that very limited spectroscopic information is contrasted by a vast variety of possible carriers. In order to reduce the number of possibilities, an increasing knowledge about the conditions in the sources, their molecular composition and in particular highly precise predictions from *ab initio* calculations are desirable. This is especially the case since many of the molecules identified so far are highly reactive species such as radicals, carbenes or molecular ions, so that the measurement of spectroscopic parameters in the laboratory is very challenging.[1]

This thesis is intended to predict accurate spectroscopic parameters for l-C_3H^+ and C_4. Both molecules are of interest to interstellar cloud chemistry and only scarce information about their rovibrational properties is available. Previous spectroscopic and theoretical work will be summarized in the following sections.

1.1 Previous experimental and theoretical results for l-C_3H^+

The cation l-C_3H^+ is of great interest to astrochemistry where it is assumed to play an important role in ion-molecule reactions producing longer carbon-chain containing molecules.[2,3] Its reaction with hydrogen is further believed to be a key step in models of the hydrocarbon chemistry in molecular clouds.[4,5]

Although its importance has been discussed, this cation was only recently detected in the interstellar medium (ISM). In 2012, Pety *et al.* published the detection of eight harmonically related radio lines in their survey of the Horsehead photodissociation region, which they attributed to l-C_3H^+.[6] They deduced a rotational constant in the vibrational ground state B_0 of 11244.9512 ± 0.0015 MHz. Since no experimental rotational spectrum was available at that time, the assignment was based on previous *ab initio* calculations that predicted a linear structure with a $^1\Sigma$ electronic state and a rotational constant of about 11.1 GHz.[7] Later, McGuire *et al.* re-examined previously published and publicly available data from observations toward Sagittarius B2(N), Sagittarius B2(OH) and the dark cloud TMC-1 and found evidence for the observed lines in the former two sources as well as trace evidence in the latter source.[8] This assignment was then questioned by Huang *et al.* who calculated a complete quartic force-field (QFF) on the basis of coupled cluster with singles, doubles and perturbative triples (CCSD(T)) calculations with correlation consistent basis sets up to cc-pV5Z and an extrapolation to the complete basis set (CBS) limit of these energies via a three-point formula. This QFF was further corrected for scalar relativity and core correlation.[9] The geometry and spectroscopic parameters obtained from this force field were in very good agreement with the B_0 value derived from the ISM measurements but the calculated centrifugal distortion constant at equilibrium D_e differed by about 45 % from the experimental D_0 value. This discrepancy led the authors to the conclusion that the experimental assignment was wrong. In a subsequent paper by the same group it was proposed that the kinetically stable quasi-linear isomer of l-C_3H^- is the carrier of the observed lines.[10] This reassignment, however, was rejected by McGuire et al. because there was no evidence for $K_a = 1$ lines and the existence of this anion would mean the highest anion-neutral ratio yet observed in the ISM.[11] Only months later, Brünken et al. published a laboratory rotational spectrum of l-C_3H^+ from a recently invented mass-selective action spectroscopy method in a 4 K cryogenic ion trap apparatus.[12] A fit of the four observed lines yielded perfect agreement with the astronomical data and thus confirmed the initial assignment.

Prior to the study of Huang et al., only Botschwina et al. published results obtained with vibrational perturbation theory, based on an anharmonic potential calculated by CCSD(T) with a basis set consisting of 99 contracted Gaussian type orbitals.[1] It is therefore the objective of this thesis to bring theory into agreement with the spectroscopic data available and make accurate predictions for future spectroscopic work.[1]

1.2   Previous experimental and theoretical results for C_4

Linear carbon chains C_n (n: odd) have been detected in the ISM (C_3, Ref. [14]), in diffuse interstellar clouds (C_3, Ref. [15]) as well as in carbon rich stars (C_5, Ref. [16]). It can be assumed that carbon chains with an even number of carbon atoms are also present in some of the sources noted above and are important for reactions leading to complex carbon-rich species.[17] Although not yet unambiguously detected, there has been a tentative assignment of C_4 based on the detection of a molecular band at 174 cm^{-1} in carbon-rich evolved stars and in Sgr B2 which was attributed to the IR-active cis bending mode (ν_5) of C_4.[17] Nevertheless, there is some doubt about the assignment because the band shape can only be accurately reproduced assuming a rather large spin-orbit interaction constant of $A_{SO} > 2$ cm^{-1}. This in turn would require an excited electronic state to lie very close to the ground electronic state. Experimental excitation energies for low lying electronic states determined by Xu et al. from photoelectron spectroscopy are 0.33 eV for the first excited state ($^1\Delta_g$) and 0.50 eV for the second excited state ($^1\Sigma_g^+$) and are thus probably too high in energy to give rise to such a large spin-orbit interaction constant.[18]

C_4 however has been studied intensively in the laboratory by argon matrix isolation spectroscopy[19,20] and infrared diode laser spectroscopy.[21–24] Several isotopomers were measured by the former method and $\nu_3 = 1543.4$ cm^{-1} and $\nu_5 = 172.4$ cm^{-1} were determined for the antisymmetric stretching wavenumber[19] and the cis bending mode[20] of the most abundant isotopologue, respectively. In case of the bending mode, a splitting of about 2 cm^{-1} was observed and attributed to different trapping sites of the matrix.

---

[1] A great part of the results for l-C_3H^+ of the thesis has already been published in The Astrophysical Journal (see Ref. [13]).

The results of the gas phase infrared (IR) diode laser experiments are the most reliable but are limited to the IR-active antisymmetric ν_3 stretching band at 1548.6128(4) cm^{-1}[23] and hot bands, where the final states are combination tones of ν_3 with one of the bending vibrations. A value of 4979.92(20) MHz was determined for B_0 and a value of 0.855(44) kHz for the corresponding quartic centrifugal distortion constant D_0.[23] The authors also determined the vibration rotation interaction constants $\alpha_3 = 21.13$ MHz, $\alpha_4 = -20.86$ MHz and $\alpha_5 = -22.5$ MHz.[22,23] From the l-type doubling constants $q_4 = 5.52(19)$ MHz and $q_5 = 10.96(12)$ MHz[23] (uncertainties of one standard deviation as obtained from the least-squares fit are given in parentheses) it was further possible to estimate the bending fundamentals with the formula

$$\nu_i = f_i^q B_0^2 / q_i \qquad . \tag{1}$$

The factor f_i^q was determined via a fit of known data for HCCCCH and NCCN. It is questionable whether these constants are reliable, given the few data points and whether the reference molecules resemble the situation in linear C_4, where the bending frequencies are much lower and the bonds are cumulenic as opposed to conjugate triple-bonded systems in the reference molecules. The error bars for the fundamentals calculated in that manner ($\nu_4 = 352(15)$ cm^{-1} and $\nu_5 = 160(4)$ cm^{-1}) are thus probably too optimistic.

In addition to these rovibrational studies, Coulomb explosion experiments[25] and measurements of the electron photodetachment cross section[26] gave indirect evidence for the presence of the cyclic isomer of C_4, indicating that this isomer might also be present in the ISM and C-rich evolved stars.

The complicated electronic structure of this molecule has raised great interest among theoretical chemists. The first quantum-chemical studies were reported in 1959 by Pitzer and Clementi who proposed a cumulenic structure with a $^3\Sigma_g^-$ ground state for the linear isomer based on semi-empirical calculations[27] and later on minimum basis self-consistent field (SCF) calculations.[28] Bartlett *et al.* were the first to publish both harmonic vibrational wavenumbers and intensities based on both unrestricted Hartree-Fock (UHF) and restricted open-shell Hartree-Fock (ROHF) calculations with a rather small 4s2p1d basis set.[29] Four years later Watts *et al.* improved on these calculations by correlating all electrons in a CCSD(T) calculation with a correlation consistent pVTZ basis set.[2] Martin and Taylor obtained the geometry and harmonic vibrational wavenumbers of the stretching modes from UHF based CCSD(T) calculations employing a rather small pVDZ basis set.[30] In 1997 Botschwina derived an accurate equilibrium geometry from CCSD-T and CCSD(T) calculations with a pCVQZ basis set and all electrons being correlated.[31] The rotational constant ($B_e = 4967.6$ MHz) derived from the geometry obtained by the latter method ($R_{in} = 1.2907$ Å and $R_{out} = 1.3098$ Å for the middle and the two outer CC

---

[2] A valence only basis set was used although all electrons were correlated.

bonds, respectively) was in somewhat fortuitous agreement with the experimental ground
state value, given that B_e is usually very close to B_0 for semi-rigid linear molecules. The
harmonic vibrational wavenumbers obtained for the totally symmetric stretching modes
($\omega_1 = 2109$ cm^{-1} and $\omega_2 = 926$ cm^{-1}) are still the best available in the literature. In
a study concerning the IR-active bending modes of linear carbon chains with an even
number of carbon atoms, Botschwina published the harmonic vibrational wavenumber of
the cis bending mode along with a calculated intensity ($\omega_5 = 171.1$ cm^{-1}; $I = 44.5$ km
mol^{-1}).[32]

Based on their previous study on the electronic structure of several isomers of C_4,[33] Senent
et al. carried out the most extensive study so far on the rovibrational spectroscopy of
the linear and rhombic isomers of C_4.[34] They computed 1050 non-redundant single-point
energies with MRCI(+Q)/VTZ and fitted these energies to a polynomial representation
of the PES in symmetry coordinates that allowed for the definition of a full QFF. The
root mean square error of 43 cm^{-1} and the estimated accuracy of only 0.01 Å for the bond
lengths raises some doubts about the accuracy of the derived spectroscopic constants. The
calculated value for the ν_3 fundamental is 52 cm^{-1} lower than the accurate experimental
value. This and some unusually large anharmonic constants (e.g. $x_{13} = -109.590$ cm^{-1})
reveal severe numerical inaccuracies. Some possible sources of error will be discussed in
the following sections.

It is the aim of this thesis to improve on the calculations published so far by calculating
a highly accurate equilibrium structure and full QFFs that are capable to obtain spectroscopic parameters that reproduce the precise experimental values available. Furthermore,
predictions on some properties that have not been measured yet will be made. It is
therefore necessary to investigate the performance of several electronic structure methods
based on both single- and multi-configurational calculations. This thesis may then also
be of interest for the study of the rovibrational spectroscopy of other molecules with an
open-shell ground state.

As will be discussed later, both a variational and a perturbational approach can be
chosen for the rovibrational calculations. Although variational calculations are without
doubt more precise, only results obtained with vibrational perturbation theory will be
presented here since a lot of different methods and the influence of several smaller effects
had to be tested. Variational calculations are much more time-consuming and require a
well determined PES to optimize the basis used in these calculations which makes fre-
quent changes in the PES or parts thereof unfeasible. Especially C_4 with its challenging
electronic structure requires the comparison of several electronic structure methods and
it is desirable to quickly analyze the effects of changes in the PES on the resulting spectroscopic parameters. The results of this study however will enable to construct accurate
PESs for both molecules to be used in variational calculations.

As stated above, a lot of flexibility was desired and it was decided to write a new vibrational perturbation theory program that enables to quickly perform the calculations and e.g. allows for the use of force constants of different order as obtained different electronic structure methods. The resulting program **4Lin** is discussed in the appendix while the underlying theory is described in section 2.2.

2 Theoretical methods

2.1 Electronic structure methods

An accurate description of the electronic structure of a molecule needs to take into account the correlation energy, which is defined as the difference between the exact energy and the Hartree-Fock (HF) energy. The choice of the method applied to calculate the correlation energy depends on the nature of the correlation. There are in general two limiting cases:

1. **dynamic correlation:** The HF determinant is a valid approximation and all calculations of the electron correlation are based on excitations from this determinant.

2. **static correlation:** The molecule has to be described by more than one determinant and the description with only one HF determinant is qualitatively wrong.

As might be expected, most molecules are somewhere in between these limiting cases. While all methods that start from the HF determinant are referred to as single reference methods, methods that start from a description with more than one determinant are called multi-reference methods. For both types of methods, the ones applied in this thesis will be briefly presented in the following.

All electronic structure calculations have been carried out with the Molpro package of *ab initio* programs[35] unless otherwise stated.

2.1.1 Single reference methods

Coupled-Cluster Calculations

As noted above, the electron correlation in single-reference methods is calculated by generating new terms in the wave function via excitations of electrons from the HF determinant. Depending on how many electrons are excited with respect to the HF determinant, these new determinants are classified as *single, double, triple* etc. excitations. If an excitation operator $\hat{T}$ is defined as:[36]

$$\hat{T} = \hat{T}_1 + \hat{T}_2 + \hat{T}_3 + ... + \hat{T}_{N_{el}} \tag{2}$$

this operator will create all possible excitations from the HF determinant and give the correct wave function within the given basis set. However, the number of terms is extremely large even for systems with rather few electrons. In coupled cluster (CC) theory, the excitation operator is the argument of the exponential function which accounts for the fact that there are dependencies between the coefficients of the wave function expansion for different excitations. The coupled cluster wave function is then given as:

$$\Psi_{CC} = e^{\hat{T}} \Phi_{HF} \tag{3}$$

The first time that the CC ansatz was introduced into theoretical chemistry was by Čížek in 1966[37] while an equivalent approach was used almost ten years earlier in nuclear physics.[38] Since the number of terms generated by the cluster operator would be too large to handle computationally, the excitation operator is usually truncated. In coupled cluster with singles and doubles (CCSD) for example, the operator is truncated after the $\hat{T}_2$ term. The exponential form of the cluster operator however ensures that also disconnected excitations of higher order are included, that result from products of the excitation operator eq. 2 (e.g. disconnected quadruples via the $1/2\hat{T}_2^2$ operator). This inclusion of products of excitations makes the CC method size extensive while it is not variational in the commonly used projected form.[39]

One of the most commonly used *ab initio* methods is CCSD(T),[40] where the triple excitations are included in a perturbative manner. In the case of open-shell electronic states, either restricted or unrestricted HF reference wave functions can be used. The former method leads to both a spin unrestricted (RHF-UCCSD) and a partially spin-restricted (RHF-RCCSD) formalism. RHF-UCCSD is computationally about three times more demanding than RHF-RCCSD, which is furthermore free from spin contamination.[41,42] Since no UHF based CC variants are available in Molpro, the interface to Kallay's MRCC program[43] was used for the UHF-UCCSD(T) calculations. Usually, only valence electrons are correlated in the CC calculations and results obtained with this method will be marked with the suffix fc (frozen-core), whereas calculations with all electrons being correlated will be denoted with ae (all electron).

Explicitly correlated methods

Since convergence of the correlation energy with the basis set is quite slow and CCSD(T) already has a rather unfavourable scaling of N_{basis}^7 with the basis set size, methods are required that enable to reach the CBS limit faster. One of the reasons of the slow convergence is the poor description of the electronic cusp when the wave function is expanded in functions that do not explicitly include the interelectronic distance. Hylleraas was the first to use functions that do include the interelectronic distance r_{12} (so called geminals) in 1929[44] and later that idea was revived by Kutzelnigg and co-workers.[45–47] Following a suggestion by Ten-no,[48] Werner and coworkers implemented geminals of the form $F(r_{12}) = -\frac{1}{\beta}\exp(-\beta r_{12})$ in coupled-cluster calculations.[49,50] The resulting method was presented with two different approximations with acronyms CCSD(T)-F12a and CCSD(T)-F12b, where the latter was recommended for larger basis sets and will be used in this thesis. The derivation of the theory leads to many three- and four-electron integrals whose calculation is very expensive. One possible solution is to approximate these integrals by sums of two-electron integrals by introduction of the resolution of the identity (RI).[45] An efficient implementation then requires auxiliary basis functions, the construction of a complementary auxiliary orbital basis (CABS) and density fitting (DF) approximations.[49,50] Since the treatment of the perturbative triples is not straightforward in the F12 framework, they are determined conventionally which leads to a small basis set error. Werner and coworkers proposed a simple scheme to correct for that error by scaling the triples contribution[50] according to

$$\Delta E_{(T^*)} = \Delta E_{(T)} \cdot \frac{E_{corr}^{MP2-F12}}{E_{corr}^{MP2}} \qquad (4)$$

where $E_{corr}^{MP2-F12}$ and E_{corr}^{MP2} denote the correlation energy as obtained by Møller-Plesset perturbation theory of second order with and without F12 approximation, respectively while $\Delta E_{(T)}$ is the triples contribution and the scaled triples contribution is designated by the asterisk in $\Delta E_{(T^*)}$.

2.1.2 Multi-reference methods

In typical HF-SCF calculations only the occupied orbitals of a single determinant are optimized. If, however, static correlation is present, a single determinant may lead to even qualitatively wrong results and it is necessary to expand the wave function as a linear combination of several configuration state functions (CSFs) and optimize both the orbitals and the expansion coefficients.

$$\Psi = \sum_I C_I \Phi_I \qquad (5)$$

This method is referred to as multi-configuration self-consistent field (MCSCF). The resulting equation for the energy is of fourth order in the orbitals making a direct minimization impracticable. Instead, an approximative expression for the energy is optimized iteratively (microiterations) resulting in better orbitals and coefficients leading to faster convergence.[51] The implementation of this second-order optimization in Molpro by Werner and Knowles[52,53] allows for a reduction of computational time by a factor of 10-20 compared to common Newton-Raphson procedures. In MCSCF calculations, the choice of the active space i.e. the orbitals from which excitations are considered is crucial, since an active space that is too small will only give a small fraction of the correlation energy and large active spaces can become computationally unfeasible. Orbitals that are not part of the active space will have occupation numbers of 2 or 0, whereas orbitals belonging to the active space will usually have fractional occupation numbers. If MCSCF is performed as a complete active space self-consistent field (CASSCF) calculation this means that within the active space all possible excitations are considered. CASSCF calculations can thus be considered as a full configuration interaction (CI) within the active space and in this thesis only this type of calculations will be performed.

Multi-reference configuration interaction

Analogous to CI expansions in single-reference methods it is possible to calculate dynamical correlation by including excitations from the MCSCF reference function. This leads to the multi-reference configuration interaction (MRCI) method, which is usually truncated after the double excitations and is then called multi-reference configuration interaction with singles and doubles (MRCISD). The wave function can then be written as

$$\Psi = \sum_I C_I \Phi_I + \sum_s \sum_a C_a^s \Phi_s^a + \sum_{ij,p} \sum_{ab} C_{ab}^{ij,p} \Phi_{ij,p}^{ab} , \tag{6}$$

where the configuration space is subdivided in three subspaces. Φ_I includes all internal configuration state function (CSF)s that can be constructed from all orbitals that are occupied in the reference wave function. The singly external CSFs Φ_s^a are created by single excitation from the reference. Finally, in $\Phi_{ij,p}^{ab}$ two electrons are excited to external orbitals and the index p denotes the spin state which is either singlet or triplet. As first suggested by Meyer[54] and later implemented by Werner and Reinsch,[55] the number of excitations can considerably be reduced if the excitation operators work on the complete reference wave function instead of working on the CSFs. This method of "internal contraction" is now commonly used and implemented in Molpro.[56,57]

Being a truncated CI method, MRCISD is not size-extensive. Several corrections have been proposed to approximately account for higher order excitations. The most popular one by Davidson[58] can be formulated as

$$\Delta E_D = E_{\text{corr}} \frac{1 - c_0^2}{c_0^2} , \tag{7}$$

where c_0 is the coefficient of the reference wave function in the MRCI wave function and E_{corr} is the correlation energy as obtained by MRCI. It is possible to either use the coefficient of the fixed or the relaxed reference function, but the latter choice is made throughout this thesis. When the Davidson correction is employed, the acronym is MRCI(+Q) and MRCI otherwise.

Average coupled-pair functional

Another way to capture dynamical correlation and approximately achieve size extensivity is to minimize the energy functional:

$$E = E_0 + \frac{\langle \Psi_0 + \Psi_c | H - E_0 | \Psi_0 + \Psi_c \rangle}{1 + g_a \langle \Psi_a | \Psi_a \rangle + g_e \langle \Psi_e | \Psi_e \rangle} \tag{8}$$

where Ψ_0 and Ψ_c are the normalized reference and correlation wave functions, respectively, Ψ_a and Ψ_e are the internal and external parts of Ψ_c and g_a and g_e are numerical factors.

The choice of $g_a = 1$ and $g_e = 2/n$ with n being the number of correlated electrons leads to the size extensive average coupled-pair functional (ACPF) method.[59] If $g_a = g_e = 1$, the usual MRCI solution is retained. For both MRCI and ACPF, an explicitly correlated method has been implemented in Molpro, which uses similar approximations as the F12b variant for single-reference coupled cluster.[60]

CIPT2

In order to also include correlation effects from the core electrons, a hybrid method, CIPT2,[61] where excitations from the active space are treated by MRCI and the remaining excitations of the core electrons are treated by multi-reference perturbation theory at second-order (MRPT2), was developed. This has the advantage of including core correlation effects at rather low cost without introducing intruder-state problems that are an issue for conventional MRPT2 calculations. Since these problems usually occur due to excitations from the active orbital subspace which is now treated by MRCI, they are effectively reduced as was shown for calculations on the chromium dimer.[61] As is the case for MRCI, the method can be employed with and without the use of Davidson correction, leading to acronyms CIPT2(+Q) and CIPT2, respectively.

For all methods applied in this thesis, correlation consistent polarized valence basis sets of the Dunning type are employed.[62] The acronym cc-pVnZ with n taking the values T,Q,5,6,7 will be further abbreviated to VnZ in the following. For the explicitly correlated methods, basis sets of the cc-pVnZ-F12 type (briefly termed VnZ-F12) are employed.[63] For the RI approximation, auxiliary basis sets of the cc-pVnZ-F12/Optri[64] type where n is the same for the atomic-orbital and the auxiliary basis set are employed. Furthermore basis sets of the aug-cc-pVnZ/MP2FIT[65] and the cc-pVnZ/JKFIT[66] type are used as DF basis and DF basis for the Fock and exchange matrices, respectively.

2.1.3   Inclusion of smaller effects

The methods described so far build the basis of a composite approach where especially in the case of single-reference methods some smaller effects need to be added in order to get a larger fraction of the correlation energy.

Douglas-Kroll-Hess method

The Hamiltonians employed in all of the above methods are non-relativistic. Building up on the work of Douglas and Kroll,[67] Hess suggested the use of a transformed relativistic Hamiltonian that is bounded from below, thereby enabling variational calculations.[68,69] The Douglas-Kroll-Hess method at second-order (DKH2) is then capable of taking into account scalar-relativistic effects and typically recovers more than 97 % of the

total relativistic energy as obtained by the fully relativistic Dirac-Hartree-Fock method
in the case of atoms. The application of this method requires a specially recontracted
basis set that allows for a different radial behaviour in the core region. These basis sets
have been provided by de Jong and coworkers and are denoted VnZ-DK, using a similar
notation as above.[70]

The contribution of scalar relativistic effects as obtained by the DKH2 method will be
denoted ΔDK.

Core-core/core-valence correlation

Both the fc-CC and the multi-reference methods (excluding CIPT2) include only excitations of the valence electrons while the core electrons are still uncorrelated. The correction
of this deficiency requires the use of a different class of basis sets, which allow for more
flexibility in the core region and is denoted cc-pCVnZ (briefly termed CVnZ).[71] These
basis sets are also used in the few ae-CCSD(T) calculations in this thesis. The contribution of inner-shell correlation calculated as the difference between ae- and fc-CCSD(T)
calculations with the same CVnZ basis set will be termed ΔCV.

Higher-order correlation

A further correction to the CC calculations can be obtained by including excitation
operators higher than $\hat{T}_2$ in eq. 2. This leads to a hierarchy of CC methods:

$$\mathbf{CCSD(T)} \rightarrow CCSDT \rightarrow \mathbf{CCSDT(Q)} \rightarrow \mathbf{CCSDTQ}$$

where the brackets denote a perturbative treatment of the respective excitation operator
and only the methods printed in bold are considered in this thesis. The reason for the
neglect of explicitly calculating the effect of the iterative triples contribution is that usually
an overestimation of the correlation energy is observed and CCSD(T) performs better than
CCSDT in most cases when applied to e.g. spectroscopic parameters.[72] The contributions
of higher correlation are calculated as differences. Δ(Q)-(T) denotes the effect of iterative
triples and is caluated as the difference between a fc-CCSDT(Q) and a fc-CCSD(T)
calculation employing the same basis set. ΔQ-(Q) is the difference of a fc-CCSDTQ and
a fc-CCSDT(Q) calculation. The sum of both is termed ΔHC. The CCSDT(Q) method[73]
will be applied with a VTZ basis set while the rather small VDZ basis set is employed
for the expensive CCSDTQ calculations.[74] All higher-order correlation (HC) calculations
were performed in Molpro with an interface to Kallay's MRCC program[43].

Diagonal Born-Oppenheimer correction

An improvement over the Born-Oppenheimer approximation is achieved by means of the diagonal Born-Oppenheimer correction (DBOC).[75] The energy contribution is given by:

$$\Delta E_{\text{DBOC}} = -\sum_{A=1}^{N} \sum_{i=x,y,z} \frac{1}{2M_A} \langle \Psi(\mathbf{r};\mathbf{R}) | \nabla_{R_{A_i}}^2 | \Psi(\mathbf{r};\mathbf{R}) \rangle \qquad (9)$$

It is thus calculated as the expectation value of the electronic Born-Oppenheimer wave function $\Psi(\mathbf{r};\mathbf{R})$ over the sum of the kinetic energy operators of the nuclei $\nabla_{R_{A_i}}^2$ with M_A denoting the masses of the nuclei, $\mathbf{r}$ and $\mathbf{R}$ are coordinates of the electrons and the nuclei, respectively. Eq. 9 shows that the DBOC correction is most important for light nuclei. In this thesis it is considered for coordinates involving H-atom motion only. The DBOC contribution (termed ΔDBOC) and the corresponding $\Psi(\mathbf{r};\mathbf{R})$ wave function is calculated by means of ae-CCSD with a CVQZ basis set using the CFOUR program.[76]

2.2 Vibrational Perturbation Theory

Watson's isomorphic rovibrational Hamiltonian for linear molecules has the form:[77]

$$\hat{H} = \underbrace{\frac{1}{2}\sum_{k=1}^{3N-5} P_k^2 + \frac{1}{2}\mu(\pi_x^2 + \pi_y^2)}_{\hat{H}_{\text{vib}}} + \underbrace{\frac{1}{2}\mu(\Pi_x'^2 + \Pi_y'^2)}_{\hat{H}_{\text{rot}}} - \underbrace{\mu(\Pi_x'\pi_x + \Pi_y'\pi_y)}_{\hat{H}_{\text{Coriolis}}} + V\,, \qquad (10)$$

where P_k are linear vibrational momenta conjugate to the normal coordinates Q_k, $\pi_{x,y}$ and $\Pi_{x,y}'$ are vibrational and rotational angular momenta, respectively, μ denotes the inverse inertia tensor and V is the PES represented in normal coordinates. As is the case in electronic structure calculations, there are two general ways two find the eigenstates of this Hamiltonian. The variational approach of diagonalizing the Hamiltonian in a suitable basis (e.g. products of harmonic oscillator/rigid rotor functions) or perturbation theory to arbitrary order. The former approach gives more accurate results and has widely been applied in our group on 3-atomic molecules.[78,79] The calculations, however, become increasingly more complex with the number of atoms especially since the rovibrational basis required becomes very large. Therefore perturbation theory is commonly used for larger molecules, yielding spectroscopic parameters in reasonable agreement with experimental data in the case of semi-rigid molecules.[30,80] Apart from model calculations, only vibrational perturbation theory at second-order (VPT2) will be applied in this thesis and discussed in more detail in the following.

When calculating the second order perturbation energy, a large number of off-diagonal matrix elements of the first-order perturbation Hamiltonian need to be evaluated. A contact transformation ensures that all these elements are zero, thereby formally reducing the problem to a first-order calculation. This is achieved by a similarity transformation

$$\hat{H}' = T\hat{H}T^{-1} \tag{11}$$

where the unitary function T takes the form:

$$T = e^{i\lambda \mathbf{S}} \quad . \tag{12}$$

T can now be expanded in powers of λ and insertion into eq. 11 and collecting same powers of λ ultimately leads to

$$\hat{H}^{(1)} = i(\hat{H}^{(0)}S - S\hat{H}^{(0)}) \tag{13}$$

when applying the condition $\hat{H}^{(1)'} = 0$. The form of the transformation function S has been discussed by Herman and Shaffer[81] and Nielsen.[82]

The second order transformed Hamiltonian is then given by

$$\hat{H}^{(2)'} = \sum_k \hat{H}_k^{(2)} + \sum_{tr} \frac{i}{2}[S_r\hat{H}_t^{(1)} - \hat{H}^{(1)}S_r] \tag{14}$$

where the second sum is only over terms of second-order.[83] The remaining diagonal matrix elements were computed and are tabled in the article by Herman and Shaffer.[81] By separately collecting terms with like powers of vibrational and rotational quantum numbers the term energy formulas:[84]

$$T(v, J) = G(v) + F_v(J) \tag{15}$$

with

$$G(v) = \sum_r \omega_r(v_r + 0.5d_r) + \sum_{r \geq s} \chi_{rs}(v_r + 0.5d_r)(v_s + 0.5d_s) + \sum_{t \geq t'} \chi_{l_t l_{t'}} l_t l_{t'} + \dots \tag{16}$$

and

$$F_v(J) = B_v(J(J+1) - l^2) - D_e(J(J+1) - l^2)^2 + H_e(J(J+1) - l^2)^3 + \dots \tag{17}$$

are obtained. In eq. 16 ω_r denotes the harmonic, χ_{rs} and $\chi_{l_t l_{t'}}$ denote anharmonic vibrational constants, d_r is the degeneracy of the r-th normal mode and the index t sums over degenerate normal modes only while r and s sum over all normal modes. Finally, v is a collective variable for the set of all vibrational quantum numbers. B_v, D_e and H_e

in eq. 17 are the effective rotational constant, quartic and sextic centrifugal distortion constants, respectively.

Each of these spectroscopic parameters is now connected to the potential (see eq. 10). The following formulas as taken over from Allen *et al.*[84] require the anharmonic force constants of the potential expanded in dimensionless normal coordinates using the unrestricted summation convention.[3] The vibrational anharmonic constants are then given by:

$$\chi_{ss} = \frac{1}{16}\phi_{ssss} - \frac{1}{16}\sum_{s'}\phi_{sss'}^2 \frac{8\omega_s^2 - 3\omega_{s'}^2}{\omega_{s'}(4\omega_s^2 - \omega_{s'}^2)}, \tag{18}$$

$$\chi_{tt} = \frac{1}{16}\phi_{tttt} - \frac{1}{16}\sum_{s}\phi_{stt}^2 \frac{8\omega_t^2 - 3\omega_s^2}{\omega_s(4\omega_t^2 - \omega_s^2)} \tag{19}$$

$$\chi_{l_t l_t} = -\frac{1}{48}\phi_{tttt} - \frac{1}{16}\sum_{s}\phi_{stt}^2 \frac{\omega_s}{4\omega_t^2 - \omega_s^2}, \tag{20}$$

in case of the diagonal elements and

$$\chi_{ss'} = \frac{1}{4}\phi_{sss's'} - \frac{1}{4}\sum_{s''}\phi_{sss''}\phi_{s''s's'}\frac{1}{\omega_{s''}} - \frac{1}{2}\sum_{s''}\phi_{ss's''}^2 \frac{\omega_{s''}(\omega_{s''}^2 - \omega_s^2 - \omega_{s'}^2)}{\Delta_{ss's''}}, \tag{21}$$

$$\chi_{st} = \frac{1}{4}\phi_{sstt} - \frac{1}{4}\sum_{s'}\phi_{sss'}\phi_{s'tt}\frac{1}{\omega_{s'}} - \frac{1}{2}\sum_{t'}\phi_{stt'}^2 \frac{\omega_{t'}(\omega_{t'}^2 - \omega_s^2 - \omega_t^2)}{\Delta_{stt'}} + B_e\zeta_{st}^2\left(\frac{\omega_s}{\omega_t} + \frac{\omega_t}{\omega_s}\right), \tag{22}$$

$$\chi_{tt'} = \frac{1}{8}(\phi_{t_x t_x t'_x t'_x} + \phi_{t_x t_x t'_y t'_y}) - \frac{1}{4}\sum_{s}\phi_{stt}\phi_{st't'}\frac{1}{\omega_s} - \frac{1}{4}\sum_{s}\phi_{stt'}^2 \frac{\omega_s(\omega_s^2 - \omega_t^2 - \omega_{t'}^2)}{\Delta_{stt'}}, \tag{23}$$

$$\chi_{l_t l_{t'}} = \frac{1}{2}\sum_{s}\phi_{stt'}^2 \frac{\omega_s\omega_t\omega_{t'}}{\Delta_{stt'}} \tag{24}$$

in case of the off-diagonal elements. The index s runs only over non-degenerate modes and the index t runs only over degenerate modes while the denominator $\Delta_{rr'r''}$ is given by

$$\Delta_{rr'r''} = (\omega_r + \omega_{r'} + \omega_{r''})(\omega_r + \omega_{r'} - \omega_{r''})(\omega_r - \omega_{r'} + \omega_{r''})(\omega_r - \omega_{r'} - \omega_{r''}). \tag{25}$$

B_e is the rotational constant at equilibrium and ζ_{st} are the Coriolis interaction constants whose calculation is described in the appendix.

The effective rotational constant B_v in eq. 17 is given by

$$B_v = B_e - \underbrace{\sum_{r}\alpha_r(v_r + 0.5d_r)}_{-\Delta B_0} + ... \tag{26}$$

---

[3]Note that this choice results in formulas that differ from the ones originally obtained by Nielsen[82,85] by multiplicative factors.

and analogous formulas hold for the centrifugal distortion constants. In eq. 26, α_r are the vibration-rotation interaction constants with respect to the r-th normal mode and read

$$\alpha_s = -\frac{2B_e^2}{\omega_s}\left[\frac{3a_s^2}{4I_e} + \sum_t \zeta_{st}^2 \frac{3\omega_s^2 + \omega_t^2}{\omega_s^2 - \omega_t^2} + \pi\left(\frac{c}{h}\right)^{1/2}\sum_{s'}\phi_{sss'}a_{s'}\frac{\omega_s}{\omega_{s'}^{3/2}}\right] \tag{27}$$

$$\alpha_t = -\frac{2B_e^2}{\omega_t}\left[\frac{1}{2}\sum_s \zeta_{st}^2 \frac{3\omega_t^2 + \omega_s^2}{\omega_t^2 - \omega_s^2} + \pi\left(\frac{c}{h}\right)^{1/2}\sum_s\phi_{stt}a_s\frac{\omega_t}{\omega_s^{3/2}}\right]. \tag{28}$$

In the above formulas I_e is the moment of inertia taken at equilibrium position and a_s are the inertial derivatives over the normal coordinates defined as:

$$a_s = (\partial I_{xx}/\partial Q_s)_e = (\partial I_{yy}/\partial Q_s)_e. \tag{29}$$

In the expansion eq. 17 the quartic centrifugal distortion constant is given by

$$D_e = \frac{1}{2}\sum_s\frac{B_s^2}{\omega_s}, \tag{30}$$

where the summation runs over totally symmetric normal modes only and the rotational derivatives B_s are defined as:

$$B_s = -\frac{\hbar^3}{2h^{3/2}c^{3/2}\omega_s^{1/2}}\frac{a_s}{I_e^2}. \tag{31}$$

The sextic centrifugal distortion constant H_e is then given by

$$H_e = \frac{4D_e^2}{B_e} - 2B_e^2\sum_s\frac{B_s^2}{\omega_s^3} - \frac{1}{6}\sum_{ss's''}\phi_{ss's''}\frac{B_sB_{s'}B_{s''}}{\omega_s\omega_{s'}\omega_{s''}}, \tag{32}$$

where the summation again runs over totally symmetric normal modes only.

Further spectroscopic parameters to be considered are the rotational and vibrational l-type doubling constants. The former can be approximated by:

$$q_t \approx q_t^e + q_t^J J(J+1) + q_t^K(K\pm1)^2 \tag{33}$$

but in this study only the harmonic part q_t^e will be calculated and is given by

$$q_t^e = -\frac{2B_e^2}{\omega_t}\sum_s\zeta_{st}^2\frac{3\omega_t^2 + \omega_s^2}{\omega_t^2 - \omega_s^2}. \tag{34}$$

The vibrational l-type doubling constants are given by

$$r_{tt'} = -\frac{1}{2}\sum_s\phi_{stt'}^2\frac{\omega_s(\omega_s^2 - \omega_t^2 - \omega_{t'}^2)}{\Delta_{stt'}} + \frac{1}{4}(\phi_{t_xt_xt'_xt'_x} - \phi_{t_xt_xt'_yt'_y}). \tag{35}$$

All of the spectroscopic parameters mentioned above are calculated by the VPT2 program **4Lin** which will be presented in the appendix.

In case of an anharmonic resonance $2\omega_a \approx \omega_b$ (Fermi type 1) or $\omega_a + \omega_b \approx \omega_c$ (Fermi type 2), some terms in the denominator of the formulas for the anharmonic constants are close to zero, making these terms unphysically large. These resonant terms are then removed and the anharmonic vibrational wavenumbers are calculated without these terms.[83,86,87] The Fermi interaction is then accounted for by setting up a 2×2 matrix with the diagonal elements given by the new anharmonic wavenumbers and the off-diagonal elements are given by the coupling elements of the two states over the first order perturbation Hamiltonian. The matrix elements for this first order treatment are tabulated e.g. in Califano p. 269 and p. 296.[83] For the most abundant isotopologues of the two molecules discussed in this thesis only one Fermi type 2 resonance occurs between the combination band $(\nu_2 + \nu_3)$ and the ν_1 stretching fundamental of l-C_3H^+. The secular determinant of the resulting matrix reads:

$$\begin{vmatrix} (\nu_2 + \nu_3) - E & \sqrt{\tfrac{1}{8}}\phi_{123} \\ \sqrt{\tfrac{1}{8}}\phi_{123} & \nu_1 - E \end{vmatrix} = 0 \tag{36}$$

leading to the equation:

$$E_{\text{up/low}} = \frac{(\nu_2 + \nu_3) + \nu_1}{2} \pm \sqrt{\left(\frac{(\nu_2 + \nu_3) + \nu_1}{2}\right)^2 - \nu_1(\nu_2 + \nu_3) + \frac{\phi_{123}^2}{8}} \tag{37}$$

for the upper E_{up} and the lower E_{low} component of this Fermi diad.

3   Results for l-C$_3$H$^+$

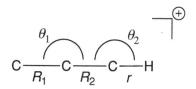

Figure 1: Specification of the internal coordinates for l-C$_3$H$^+$.

Orienting l-C$_3$H$^+$ such that the H-atom is at the right (see Figure 1), R_1, R_2 and r denote the C$_1$C$_2$, C$_2$C$_3$ and C$_3$H stretching coordinates, respectively, while θ_1 and θ_2 are the C$_1$C$_2$C$_3$ and C$_2$C$_3$H linear angle bending coordinates. The analytical representation of the potential energy function (PEF) is obtained by a polynomial fit of the calculated energies to

$$V^\alpha - V^\alpha_{\text{ref}} = \sum_{ijklm} C^\alpha_{ijklm} \Delta R^i_1 \Delta R^j_2 \Delta r^k \Delta(\sin\theta_1)^l \Delta(\sin\theta_2)^m, \tag{38}$$

where α is one of the different methods or contributions of the composite approach. In case of the single reference methods described below, the fit was performed with $\Delta\theta$ instead of $\Delta\sin(\theta)$ for the linear angle bending coordinates. This only affects two types of quartic force constants, namely $f_{\theta\theta\theta\theta}$ and $f_{\theta\theta\theta\theta'}$. These force constants however can easily be transformed to the $a = \sin(\theta)$ coordinate by the relations:[88]

$$f_{aaaa} = f_{\theta\theta\theta\theta} + 4f_{\theta\theta} \tag{39}$$

and

$$f_{aaaa'} = f_{\theta\theta\theta\theta'} + f_{\theta\theta'}. \tag{40}$$

When expressed in potential terms (denoted C_{abcd}), these formulas become

$$C_{aaaa} = C_{\theta\theta\theta\theta} + C_{\theta\theta}/3 \tag{41}$$

and

$$C_{aaaa'} = C_{\theta\theta\theta\theta'} + C_{\theta\theta'}/6. \tag{42}$$

The reference structure is chosen to be $R^{\text{ref}}_1 = 1.340$ Å, $R^{\text{ref}}_2 = 1.236$ Å, and $r^{\text{ref}} = 1.079$ Å, which is close to the equilibrium structure of the composite methods as described in the following. Both a composite scheme based on single-reference methods as well as

Table 1: Calculated equilibrium bond lengths (in Å) and equilibrium rotational constants (in MHz) for l-C_3H^+.

method	basis	$R_{1e}(C_1C_2)$	$R_{2e}(C_2C_3)$	$r_e(CH)$	B_e
fc-CCSD(T)	V5Z	1.343759	1.238593	1.080509	11115.37
fc-CCSD(T)	V6Z	1.343484	1.238322	1.080398	11119.94
fc-CCSD(T)	V7Z (no k)	1.343341	1.238206	1.080346	11122.11
fc-CCSD(T)	V8Z (no k, l)	1.343329	1.238175	1.080322	11122.49
fc-CCSD(T)-F12b	VQZ-F12	1.343176	1.238060	1.080424	11124.55
fc-CCSD(T*)-F12b	VQZ-F12	1.343307	1.238322	1.080445	11121.31
above + ΔCV		−0.003568	−0.003013	−0.001310	+55.25
above + ΔDK		−0.000180	−0.000261	−0.000149	+3.80
above + ΔHC		+0.001299	+0.000564	−0.000214	−14.98
above + ΔDBOC = **Comp. 1**		−0.000002	+0.000003	+0.000209	−0.24
		1.340856	1.235615	1.078981	11165.14
MRCI(+Q)	VQZ-F12	1.344350	1.239042	1.079934	11107.60
above + ΔCV		−0.003603	−0.003016	−0.001307	+55.42
above + ΔDK		−0.000181	−0.000261	−0.000149	+3.80
above + ΔDBOC = **Comp. 2**		−0.000002	+0.000003	+0.000209	−0.25
		1.340564	1.235768	1.078687	11166.57
Huang et al.[9]		1.339841	1.235360	1.078961	11175.51

one based on multi-reference methods is used to calculate an accurate PEF from which the cubic force-field (CFF) can be derived. The single-reference composite method is based on fc-CCSD(T*)-F12b calculations with the VQZ-F12 basis set. Subsequently, ΔCV as calculated by CCSD(T) with a CV6Z basis set, ΔDK as obtained by the VQZ-DK basis set, Δ(Q)-(T)/VTZ and ΔQ-(Q)/VDZ are added, as well as ΔDBOC with the CVQZ basis set for the CH stretching coordinate. This composite method will be referred to as **Comp. 1**.

Another PEF was constructed from MRCI(+Q)-F12 calculations with the VQZ-F12 basis set on a full-valence CASSCF reference function (all valence orbitals are in the active space). The same ΔCV, ΔDK and ΔDBOC contributions as determined from the single-reference calculations were then added since they are considered to be sufficiently method independent. This multi-reference based composite method will be denoted **Comp. 2**.

In order to decide if the triples scaling should be applied for the explicitly correlated calculations, conventional fc-CCSD(T) calculations were also performed with basis sets up to V8Z (with k and l functions being neglected). As can be seen from Table 1, both the scaled and unscaled variants agree to about $1 \cdot 10^{-4}$ Å with the largest conventional calculations in case of all bond lengths. Since the structure obtained with the scaled triples leads to a rotational constant that is closer to the conventional V7Z and V8Z values, this variant will be used as the basic contribution of the composite potential although the minimal difference makes the choice somewhat arbitrary. In the middle part of Table 1, the influence of the smaller contributions on the structure is shown for **Comp. 1**. The largest effect is due to the core-valence correlation and seems to be slightly underestimated by

Table 2: Calculated harmonic vibrational wavenumbers (in cm^{-1}) for l-C_3H^+.

method	basis	$\omega_1(\sigma)$	$\omega_2(\sigma)$	$\omega_3(\sigma)$	$\omega_4(\pi)$	$\omega_5(\pi)$
fc-CCSD(T)	V5Z	3302.8	2132.2	1182.7	801.5	124.9
fc-CCSD(T)	V6Z	3302.8	2132.9	1183.1	802.4	124.8
fc-CCSD(T)	V7Z (no k)	3303.1	2133.2	1183.3	802.1	123.4
fc-CCSD(T)	V8Z (no k, l)	3303.0	2132.9	1183.2	802.3	123.3
fc-CCSD(T)-F12b	VQZ-F12	3303.0	2134.1	1183.9	802.8	123.7
fc-CCSD(T*)-F12b	VQZ-F12	3302.5	2132.6	1183.1	802.2	123.9
Comp. 1-ΔHC		3308.4	2142.0	1189.2	806.5	126.6
Comp. 1		3308.4	2135.9	1183.0	803.1	129.6
MRCI(+Q)	VQZ-F12	3306.1	2128.1	1177.5	799.3	128.7
Comp. 2		3311.0	2137.6	1183.7	803.4	131.3
Huang *et al.*[9]		3309.7	2142.7	1189.3	805.8	124.0

Huang *et al.* (last entry of Table 1) who used a smaller basis set. When HC effects are included, R_{1e} and R_{2e} are increased by 0.001299 Å and 0.000564 Å, respectively, while r_e is slightly decreased. The resulting reduction of B_e by 15 MHz shows the necessity to include the HC effect in order to arrive at a rotational constant that is accurate to within 2-3 MHz. The effect of the DBOC reduces the CH bond length by 0.000209 Å but has almost no effect on the rotational constant since the H-atom is very light.

The lower part of Table 1 shows the effect of the smaller contributions for the multi-reference based **Comp. 2**. All three contributions decrease the bond lengths, thereby increasing B_e. It has to be noted that both composite methods lead to B_e values that differ less than 1.5 MHz, although they are based on conceptually different approaches. This confirms the reliability of the present calculations and enables to safely estimate the accuracy of the calculated B_e to be better than 3 MHz.

Calculated harmonic wavenumbers as obtained by the two methods are listed in Table 2 along with values from conventional CC calculations. Again, the values obtained from fc-CCSD(T*)-F12b agree well with those obtained by conventional fc-CCSD(T) with the V7Z and V8Z basis sets. The difference between the scaled and unscaled triples is small and not completely systematic for this quantity. The inclusion of HC effects is important for all vibrations but ω_1. This can be explained by the fact that this vibration is an almost isolated proton stretching vibration where the effects of HC are expected to be small. A further confirmation of this is the small effect on the CH bond length. In all other cases however, the HC effect is not negligible and ranges between 3 and 6 cm^{-1}.

When comparing the **Comp. 1** values with the HC effect being neglected with the values of Huang *et al.*, an excellent overall agreement is observed, the largest difference being 2.6 cm^{-1} in the case of ω_5. The reason for the latter disagreement might again be the smaller basis set used for the core-valence correlation. Although this issue can not be resolved satisfactorily, a discrepancy in the description of the lowest bending vibration should be noted. The largest difference between the value obtained by MRCI(+Q) and

Table 3: Spectroscopic parameters for l-C$_3$H$^+$ obtained from cubic force fields.

	fc-CCSD(T*)-F12b	Comp. 1-ΔHC	Comp. 1	MRCI(+Q)	Comp. 2
α_1 / MHz	32.9	33.1	32.9	32.6	32.8
α_2 / MHz	74.3	74.6	74.8	74.6	74.9
α_3 / MHz	36.9	37.0	37.9	37.6	37.8
α_4 / MHz	-3.0	-3.2	-3.1	-2.9	-3.1
α_5 / MHz	-157.2	-155.2	-151.0	-150.7	-149.0
q_4^e / MHz	13.1	13.2	13.2	13.1	13.2
q_5^e / MHz	67.6	66.9	65.2	65.0	64.4
B_e / MHz	11121.30	11180.36	11165.14	11107.60	11166.82
ΔB_0 / MHz	88.2	86.1	81.4	81.1	79.4
B_0 / MHz	11209.5	11266.4	11246.5	11180.7	11246.2
D_e / kHz	4.228	4.252	4.274	4.249	4.272
H_e / mHz	0.380	0.387	0.340	0.335	0.345

Comp. 2 is observed for the ω_2 stretching wavenumber where it amounts to 9.5 cm^{-1}, which is mainly due to ΔCV since the effect of ΔDK is quite small but in the same direction. An encouraging agreement between both composite methods is observed again.

In Table 3, spectroscopic parameters as obtained by force fields from different electronic structure methods are compared. The vibration rotation interaction constants α_i are in very good agreement for all methods in case of α_{1-4}. Only for α_5, the values differ by more than 8 MHz between the different force fields. Although this is mainly due to the effect of HC, ΔCV also decreases the absolute value by about 2 MHz as can be seen from the difference between CCSD(T*)-F12b and **Comp. 1-ΔHC** or MRCI(+Q) and **Comp. 2**. The B_0 values (cf. eq. 26) obtained by the two composite methods differ by only 0.3 MHz and are thus in excellent agreement with the radio astronomical and spectroscopic result of 11244.95 MHz.[6,12]

The l-type doubling constants $q_{4/5}^e$ as well as the quartic and sextic centrifugal distortion constants are also not very method dependent and the agreement between the two composite methods is again excellent. Given the small absolute values and the fact that even a small error in one of the three terms may lead to a wrong sign (see eq. 32), the usefulness of the sextic centrifugal distortion constant as obtained by VPT2 is questionable.

All VPT2 calculations were performed using nuclidic masses (m_H = 1.0078250 u, m_C = 12 u). Given the positive charge of l-C$_3$H$^+$, which is at least partly located at the H-atom, a slightly smaller value for m_H seems to be more appropriate with the mass of the naked proton considered to be a limiting value. This lower mass would increase ω_1 by 0.7 cm^{-1} and ω_4 by 0.2 cm^{-1}. B_e and B_0 would increase by 0.7 MHz, which is again considered to be upper limits.

Table 4: Quartic centrifugal distortion constants (in kHz) of some linear interstellar molecules.

	C_3N^-	C_3O	l-C_3H^+	C_3
D_e(theor.)	0.633[92]	0.538[93]	4.274[12]	6.077[94]
D_0(exp.)	0.686	0.777	7.709	44.099
f_D	1.08	1.44	1.80	7.26

The criticism of the original assignment of the observed harmonically related radio lines to l-C_3H^+ was based on the large difference (about 45%) between the observed D_0 and the calculated D_e value.[9] Since within VPT2, D_e is calculated from the equilibrium rotational constant and the totally symmetric harmonic wavenumbers only, it is completely independent of the flexibility of the molecule with respect to the bending coordinates. In the experimentally determined D_0 values however, the zero-point vibrational motion and therefore the effects of the bending vibrations are incorporated. As has been emphasized previously by Botschwina,[89–91] the ratio $f_D = D_0/D_e$ is connected to the floppyness of a molecule. Using the spectroscopical value of $D_0 = 7.685$ kHz[12] and the D_e value of Table 3 as obtained with either of the composite methods, the rather large value of f_D(l-C_3H^+) = 1.80 is obtained. When **Comp. 1** is applied to compute the D_e values of three other linear interstellar molecules (C_3N^-, C_3O and C_3) which contain a CCC moiety, it is possible to compare their f_D ratios with the flatness of the respective CCC bending potentials. The calculated D_e and f_D values are listed in Table 4, along with the experimental D_0 values. Figure 2 shows the correlation of the f_D ratios obtained with the flatness of the CCC bending potential.

Model calculations can be performed using a pseudo-triatomic model in which the CH group is replaced by an atom with the mass 13.0078250 u. The three-dimensional part of the **Comp. 1** PEF ($V(R_1, R_2, \theta)$) is used in variational rovibrational calculations with a computer program written by Peter Sebald.[95] B_0 and D_0 values were obtained from a least-squares fit to the lowest ten rotational energies within the vibrational ground-state. This and the corresponding D_e value from VPT2 calculations leads to a value of f_D(l-C_3H^+)$_{mod.}$ = 1.79 which is in excellent agreement with the ratio obtained above. The discrepancy observed between D_0 and D_e for l-C_3H^+ is thus to be expected and the criticism of the assignment is untenable.

The anharmonic fundamental wavenumbers were calculated with full QFFs. Due to the high computational cost and the minor importance of the off-diagonal quartic force constants for the calculation of the anharmonic constants, these were only calculated with MRCI(+Q). Nevertheless, the off-diagonal quartic MRCI(+Q) force constants can be combined with the cubic and diagonal quartic force constants obtained by the two composite methods. In addition to that, these lower order force constants as obtained by the composite methods were extended by the off-diagonal quartic force constants of

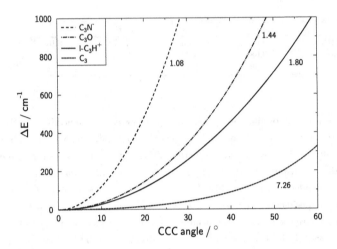

Figure 2: Plots of CCC bending potentials of some linear interstellar molecules from calculations with the **Comp. 1** method, including f_D factors.

Huang *et al.*[9] The anharmonic fundamental wavenumbers and anharmonic contributions derived from these combined force fields are listed in Table 5.

A Fermi resonance type 2 is observed for the highest stretching wavenumber because $\omega_2 + \omega_3 \approx \omega_1$. Resonant terms were then extracted from the resulting unphysically high anharmonic constants and the new anharmonic fundamental and combination tone was then calculated with first order perturbation theory as described in section 2.2. From the first two entries in the upper and lower part of Table 5 it can be concluded that the effect of the off-diagonal quartic force-constants is quite small for this molecule. That means, that an accurate description of the cubic and diagonal quartic force constants is required only in order to arrive at reliable anharmonic fundamental wavenumbers. The combinations of **Comp. 1** or **Comp. 2** with one of the available quartic force fields should thus give the best results. When comparing the effect of the off-diagonal quartic force constants as obtained by MRCI(+Q) in this thesis and those of Huang *et al.*, the difference is 1 cm^{-1} at most for the stretching vibrations and less than 2 cm^{-1} in case of the bending vibrations. The difference between **Comp. 1** and **Comp. 2** is less than 2 cm^{-1} for all vibrations. It is obvious from Table 5, that the differences in the fundamental wavenumbers are dictated by the differences in the harmonic wavenumbers obtained, since there is only little variance in the anharmonic contributions especially if all quartic force constants are considered. However, the discrepancy between the effect of quartic force constants is most pronounced for the bending vibrations. From these it can be concluded that the

Table 5: Calculated anharmonic fundamental wavenumbers (in cm^{-1}) and anharmonic contributions $\Delta_i = \nu_i - \omega_i$ (in cm^{-1}) for l-C$_3$H$^+$.

cubic:[a]	MRCI(+Q)	MRCI(+Q)	Comp. 2	Comp. 2
quartic:		MRCI(+Q)	MRCI(+Q)	Huang et al.[9]
ν_1[b]	3167.0	3165.6	3171.1	3170.1
ν_2	2080.4	2082.1	2091.8	2092.1
ν_3	1182.5	1181.1	1187.2	1187.8
ν_4	778.6	780.7	784.7	782.9
ν_5	123.7	124.6	127.9	126.1
Δ_1	-139.1	-140.4	-139.9	-140.9
Δ_2	-47.6	-46.0	-45.8	-45.5
Δ_3	5.0	3.6	3.5	4.1
Δ_4	-20.7	-18.5	-18.7	-20.5
Δ_5	-5.0	-4.1	-3.4	-5.3
cubic:[a]	Comp. 1	Comp. 1	Comp. 1	Huang et al.[9]
quartic:		MRCI(+Q)	Huang et al.[9]	Huang et al.[9]
ν_1[b]	3170.4	3169.1	3168.1	3167.8
ν_2	2088.4	2090.0	2090.3	2096.3
ν_3	1188.1	1186.7	1187.3	1194.1
ν_4	782.2	784.3	782.5	782.3
ν_5	125.3	126.1	124.3	114.2
Δ_1	-137.9	-139.3	-140.3	-141.9
Δ_2	-47.5	-45.9	-45.5	-46.4
Δ_3	5.1	3.7	4.2	4.8
Δ_4	-20.9	-18.8	-20.6	-23.5
Δ_5	-4.3	-3.5	-5.4	-9.8

[a]Here, *cubic* denotes the full cubic force field as well as diagonal quartic force constants.
[b]The Fermi resonance type 2 ($\omega_1 \approx \omega_2 + \omega_3$) has been corrected according to eq. 37.

MRCI(+Q) values are more reliable, since Huang *et al.* seem to overestimate the effect of anharmonicity as can be seen for example from their derived anharmonic contribution Δ_5 where the value is more than twice of that obtained by the different methods used in this study. A comparison of the potential terms describing the bending coordinates supports this conclusion. As can already be seen from the harmonic vibrational wavenumbers (and corresponding potential terms from Table 6) the CCC bending potential suffers from the neglect of HC correlation. While the fc-CCSD(T*)-F12b value for C_{44} is almost exactly that of Huang *et al.*, this is not the case for the value obtained with **Comp. 1**, which is more than 8 % higher. The diagonal quartic bending potential terms are almost exactly the same for both composite methods (and the corresponding basic contributions), while the value obtained by Huang *et al.* is 50 % lower in the case of C_{4444}. Furthermore, the excellent agreement of both the multi-reference and the single reference method in the case of the off-diagonal quartic bending constants increases the confidence in these values. These arguments and the systematic improvements that can be achieved by the hierarchy

Table 6: Potential terms (in atomic units) for the bending coordinates of l-C$_3$H$^+$.[a]

	Huang et al.[9]	MRCI(+Q)	fc-CCSD(T*)-F12b	Comp. 1	Comp. 2
C_{44}	0.00400	0.00433	0.00405	0.00437	0.00449
C_{55}	0.03794	0.03749	0.03777	0.03773	0.03774
C_{4444}	0.0006	0.00124	0.00113	0.00141	0.00144
C_{5555}	0.0117	0.01226	0.01251	0.01221	0.01225
C_{4455}	-0.0057	0.00117	0.00128		
C_{4445}	-0.0038	-0.00018	-0.00005		
C_{5554}	-0.0042	-0.00198	-0.00245		

[a]Coordinate 4 denotes the $\sin(\theta_1)$ CCC bending coordinate.
Coordinate 5 denotes the $\sin(\theta_2)$ CCH bending coordinate.

Table 7: Calculated spectroscopic parameters for l-C$_3$D$^+$.[a]

ω_1 / cm^{-1}	2577.38	ν_1 / cm^{-1}	2501.2	α_1 / MHz	48.4
ω_2 / cm^{-1}	2010.19	ν_2 / cm^{-1}	1968.1	α_2 / MHz	55.2
ω_3 / cm^{-1}	1157.14	ν_3 / cm^{-1}	1157.0	α_3 / MHz	30.7
ω_4 / cm^{-1}	639.52	ν_4 / cm^{-1}	626.8	α_4 / MHz	-10.7
ω_5 / cm^{-1}	123.17	ν_5 / cm^{-1}	121.0	α_5 / MHz	-127.8
B_e / MHz	10031.32				
ΔB_0 / MHz	71.3	D_e / kHz	3.265	q_4^e /MHz	13.3
B_0 / MHz	10102.6	H_e / mHz	0.295	q_5^e /MHz	55.4

[a]The cubic and diagonal quartic force constants from **Comp. 1**
and additional quartic force constants from the MRCI(+Q) force-field
were used to derive these spectroscopic parameters.

of CC methods lead to the conclusion, that the values obtained by **Comp. 1** augmented
with the off-diagonal quartic force constants from the MRCI(+Q) calculations are the best
obtained in this thesis. They should be accurate to within 2-3 cm^{-1} given that VPT2
is a good approximation for all vibrations. These values are superior to those previously
published, mostly due to the inclusion of higher correlation.

The combination of **Comp. 1** (cubic and diagonal quartic) with MRCI(+Q) (off-
diagonal quartic) can then be used to calculate spectroscopic parameters for isotopologues
of l-C$_3$H$^+$. For the deuterated isotopologue they are listed in Table 7. A Fermi resonance
for ν_3 that was indicated in the work of Huang et al.[9] was below the threshold set in
4Lin for the detection of resonances (see Appendix). The accuracy of the constants
quoted should be comparable to those of the parent isotopologue although the influence
of the DBOC has been neglected in the composite approach. The anharmonic constants
χ_{ij} and $\chi_{l_t l_t}$ are listed in Table 8. For l-C$_3$H$^+$, those neglecting the resonant terms are
given such that χ_{12} and χ_{13} are slightly too low, since the resonance is treated by first
order perturbation theory. The coupling of the ν_1 D-C/H-C stretching vibration with

the other stretching vibrations is much more pronounced in the deuterated cation since the D-C stretching is not as isolated as the H-C stretching vibration. This results in a reduction of χ_{11} by about 34 cm^{-1} while χ_{12} and χ_{13} are increased by 20 and 4 cm^{-1}, respectively.

Table 8: Calculated anharmonic constants χ_{ij} and $\chi_{l_t l_t}$ (in cm^{-1}) for l-C$_3$H$^+$ and l-C$_3$D$^+$.[a]

	$i\backslash j$	l-C$_3$H$^+$					l-C$_3$D$^+$				
		1	2	3	4	5	1	2	3	4	5
χ_{ij}	1	-54.6					-20.6				
	2	-9.3	-8.7				-29.1	-6.5			
	3	-1.1	-16.2	-2.4			-5.6	-10.8	-2.4		
	4	-18.2	-3.9	-1.8	-2.9		-13.6	-0.3	-5.1	-0.9	
	5	-2.0	-11.9	18.7	-3.2	-1.4	-4.1	-8.9	17.9	-2.4	-1.2
$\chi_{l_t l_t}$	4				5.2					1.8	
	5				1.3	1.5				1.1	1.4

[a]The cubic and diagonal quartic force constants from **Comp. 1** and additional quartic force constants from the MRCI(+Q) force-field were used to derive these anharmonic constants.

The best estimate force field can now be used to calculate intensities for both iso-topologues discussed here. Since the bending vibrations with the lower wavenumbers are not suited well for laser spectroscopy, the discussion will be limited to the stretching vibrations. The calculations were performed with a "stretch-only" Hamiltonian as implemented in Botschwina's program **linvib**,[96] which performs variational calculations over a basis of harmonic oscillator functions. Corrections due to stretch-bend coupling terms are also included,[97] the most recent application of this approach being the prediction of the absolute intensities of the totally symmetric vibrations of the propargyl cation.[98] In order to carry out these calculations, an electric dipole moment function (EDMF) was calculated around the equilibrium structure as obtained by the **Comp. 1** method. The finite field method with a field strength of ±0.0003 au along the molecular axis was applied in fc-CCSD(T*)-F12b/VQZ-F12 calculations to obtain the electric dipole moment at 87 nuclear configurations around this equilibrium structure. The calculated dipole moments were then fitted to a "stretch-only" analogue of the polynomial eq. 38 and the parameters of this EDMF are supplied in the appendix. The equilibrium dipole moment was calculated to be $\mu_e = 3.07$ D and the variation of the dipole moment with each of the bond lengths is shown in Figure 3. Integrated molar absorption intensities are then calculated with the formula

$$A_{f0} = \frac{\pi N_A}{3\hbar c_0 \epsilon_0} \nu_{f0} |\mu_{f0}|^2 \tag{43}$$

where N_A is the Avogadro constant, $\hbar$ the Planck constant divided by 2π, c_0 the speed of light in vacuum, ϵ_0 the electric constant, ν_{f0} the vibrational wavenumber and μ_{f0} the

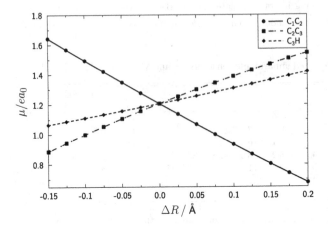

Figure 3: Variation of the electric dipole moment (in ea_0) for l-C₃H⁺ with each of the three bond lengths.

vibrational transition dipole moment for a transition from the vibrational ground state to a final state f. For stretch-bend corrections, the anharmonic constants of Table 8 were used in variational calulations and calculated intensities for both isotopologues along with the anharmonic wavenumbers from the variational calculations are listed in Table 9.

The agreement between the variational calculations and the results from VPT2 are excellent for these stretching vibrations with 2 cm⁻¹ being the largest deviation for ν_1 which has both the highest absolute value and is in Fermi resonance. However, the strength of the Fermi resonance is quite weak according to these calculations, because the contribution of the $(\nu_2 + \nu_3)$ harmonic wavefunction to the ν_1 fundamental is less than 5 %. Thus, the intensity of the combination tone $\nu_2 + \nu_3$ is very low because no intensity is "stolen" from the lower lying ν_1 fundamental which makes an experimental observation of this perturbation unlikely. The values obtained from the mechanical and electrical (double) harmonic approximation are in good agreement with the anharmonic ones.

It should be possible to detect the stretching fundamentals with IR laser spectroscopy for the most abundant isotopologue whereas the overtones and combination bands are most likely too low in intensity. Since the ν_2 fundamental has the highest intensity, it is the most promising for measurements in the IR. A technique similar to the one which led to the laboratory rotational spectrum of l-C₃H⁺[12] is available for IR spectroscopy in the range of 2560-3130 cm⁻¹.[99] With a slight extension of that energy range it should be possible to detect the P-branch in the ν_1 band of l-C₃H⁺ or the R-branch in the ν_1 band of l-C₃D⁺.

Table 9: Absolute intensities (in km/mol) and wavenumbers (in cm^{-1}) for the stretching vibrations of l-C_3H^+ and l-C_3D^+.

vibration	l-C_3H^+		l-C_3D^+	
	wavenumber	intensity[a]	wavenumber	intensity[a]
ν_3	1187.0	55.0 (55.4)	1157.1	38.0 (38.6)
ν_2	2090.7	791 (810)	1967.9	773 (794.5)
$2\nu_3$	2368.9	8.4	2309.3	4.5
ν_1	3169.1	126 (117)	2501.2	18.4 (15.2)
$\nu_2 + \nu_3$	3267.2	1.1	3114.7	9.7

[a]Values for the double harmonic approximation are given in parentheses.

4   Results for C_4 in its $X^3\Sigma_g^-$ ground state

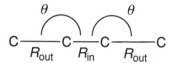

Figure 4: Specification of the internal coordinates for C_4.

The internal coordinates of the $D_{\infty h}$ symmetric molecule C_4 are specified in Figure 4. $R_{out,1/2}$ denote the two equivalent outer CC bonds while the inner CC bond is referred to as R_{in}. The two equivalent CCC linear angle bending coordinates are denoted $\theta_{1/2}$. Analogously to l-C_3H^+, the analytical representation of the PEF is obtained by a polynomial fit of the calculated energies to

$$V^\alpha - V^\alpha_{ref} = \sum_{ijklm} C^\alpha_{ijklm} \Delta R^i_{out,1} \Delta R^j_{in} \Delta R^k_{out,2} \Delta(\sin\theta_1)^l \Delta(\sin\theta_2)^m . \tag{44}$$

Due to the high symmetry, it is possible to define symmetry coordinates that allow to treat vibrations of different symmetry species separately, thereby reducing the number of energy points to be calculated. For the two totally symmetric stretching modes (σ_g) they read:

$$S_1 = \Delta R_{in} \quad \text{and} \quad S_2 = 2^{-1/2}(\Delta R_{out,1} + \Delta R_{out,2}) . \tag{45}$$

The adapted symmetry coordinate for the antisymmetric stretching vibration of σ_u symmetry is:

$$S_3 = 2^{-1/2}(\Delta R_{out,1} - \Delta R_{out,2}) \tag{46}$$

while

$$S_4 = 2^{-1/2}(\Delta\theta_1 - \Delta\theta_2) \quad \text{and} \quad S_5 = 2^{-1/2}(\Delta\theta_1 + \Delta\theta_2) \tag{47}$$

are the symmetry coordinates for the trans bending (π_g) and the cis bending (π_u) vibration, respectively. The calculations of the equilibrium structure and the harmonic vibrational wavenumbers are performed in these coordinates and only the cubic and quartic force fields are calculated in internal coordinates. Both the structure and the totally symmetric harmonic vibrational wavenumbers can be determined from a two-dimensional potential in S_1 and S_2 only. Since both Molpro and MRCC enable to use symmetry, the use of symmetry coordinates leads to a severe speed-up in the electronic structure calculations and allows for large basis sets to be applied in the case of the highly symmetric coordinates.

In the case of single-reference calculations, the open shell coupled cluster variants have to be applied because the ground state of this linear molecule is of $^3\Sigma_g^-$ symmetry. All three methods available were tested, namely partially spin-restricted ROHF-RCCSD(T), ROHF-UCCSD(T) and UHF-UCCSD(T), but only the latter two were considered in more detailed calculations. The reason for this is that severe symmetry breaking problems occured for the ROHF based variants. While the results are comparable in case of the bending and the totally symmetric modes, ω_3 as obtained with ROHF-RCCSD(T) is more than 350 cm^{-1} higher than that obtained with all other single- and multi-reference methods applied in this thesis. That the symmetry breaking problem is manifested in one (symmetry) coordinate only has also been observed by Hochlaf et al.[34] The ROHF-UCCSD(T) variant still facilitates an accurate calculation of the equilibrium structure since the coordinates involved (S_1 and S_2) are free from this symmetry breaking problem and a formalism for the treatment of HC effects with respect to that method has been derived and implemented into MRCC.[100]

Table 10: Bond lengths (in Å) and equilibrium rotational constants (in MHz) for C_4 obtained by UHF and ROHF based composite schemes.

method	UHF based		
	R_{in}	R_{out}	B_e
UCCSD(T)/CBS[4,5]	1.292085	1.311315	4956.30
$+\Delta$CV/CBS[3,4]	-0.003154	-0.003736	$+26.75$
$+\Delta$DK/CBS[3,4]	-0.000251	-0.000183	$+1.61$
$+\Delta$(Q)-(T)/VTZ	$+0.001662$	$+0.002597$	-17.01
$+\Delta$Q-(Q)/VDZ	-0.000127	-0.000414	$+2.29$
	1.290215	1.309579	4969.94
no extrapol.	1.290957	1.310758	4962.28
	ROHF based		
ae-UCCSD(T)/CV6Z	1.289794	1.309170	4973.10
$+\Delta$DK/CBS[3,4][a]	-0.000244	-0.000176	$+1.55$
$+$(Q)-(T)/VTZ	$+0.001275$	$+0.001010$	-8.52
$+$Q-(Q)/VDZ	-0.000328	-0.000023	$+1.12$
	1.290497	1.309981	4967.25

[a]Parameters were taken over from the UHF calculations.

An accurate equilibrium structure for C_4 was obtained by two composite approaches where one was based an a UHF reference while the second was based on a ROHF reference. Since the string-based algorithm for the UHF-UCCSD(T) variant as implemented in MRCC is much more time-consuming than the RHF based variants in Molpro, the calculations had to be restricted to a basis set size of V5Z and CVQZ in case of the basis contribution and ΔCV, respectively. In order to reduce the basis set incompleteness error, an extrapolation of the energy points calculated was performed with the formula:[101,102]

$$E(n) = E_{CBS} + A(n + 0.5)^{-4}. \tag{48}$$

Table 10 lists the effects on the structure and the resulting B_e value upon inclusion of smaller effects. The inclusion of core-valence (CV) correlation decreases the inner CC bond by 0.003154 Å and the outer CC bond by 0.003736 Å. The change upon inclusion of scalar-relativistic effects is quite small as usual for molecules with first row atoms only, but ΔHC leads to a decrease in the rotational constant by 14.72 MHz and is thus not negligible. From the energies obtained with the largest basis set of the respective contribution without further extrapolation, a rotational constant which is more than 7.5 MHz lower than the extrapolated value is obtained. This is due to the fact that neither the basis contribution nor ΔCV is converged with the respective basis set size and both effects lead to a decrease of the bond length. Comparing the equilibrium structure as obtained by CBS extrapolated UHF-UCCSD(T) with that obtained by UHF-UCCSD(T)/V5Z (see Table 11), a difference of 4.5 MHz is observed making the basis set incompleteness of the basis contribution the most significant. The lower part of Table 10 shows the results obtained by a ROHF based composite approach, with all electrons being correlated employing the rather large CV6Z basis set. When the B_e value of the basis contribution is compared with the combined basis+ΔCV value of the UHF based approach, a difference of 10 MHz is observed. Since in this case the effect of ΔHC amounts to only -7.40 MHz, the resulting value of $B_e = 4967.25$ MHz is in very good agreement with the UHF based approach where it has to be noted that ΔDK was taken from the upper part of Table 10. The sound overall agreement leads to a best estimate equilibrium rotational constant of $B_e = 4968 \pm 3$ MHz.

In the following, the results of several multi-reference methods will be compared to those obtained by single-reference methods. In order to apply the multi-reference methods, a suitable choice of the active space is required. While limited by the computational effort, all orbitals that change significantly with changes in the structure have to be included in the active space in order to obtain a reliable PES. The electron configuration of this molecule is given by: $(\text{core}+)3\sigma_g^2\,3\sigma_u^2\,4\sigma_g^2\,4\sigma_u^2\,5\sigma_g^2\,1\pi_u^4\,1\pi_g^2$. In the calculations carried out in this thesis, an active space was chosen where 10 electrons are distributed among 10 orbitals leading to a [10,10]-CASSCF approach. The active space includes the $4\sigma_u$ and the $5\sigma_g$ orbitals as well as the complete valence π-system ($1\pi_{u/g}$ and $2\pi_{u/g}$). Depending on the symmetry applied in the calculations, the consistency of this active space has to be ensured by proper orbital rotations. As noted above, the harmonic vibrational wavenumbers and the determination of the equilibrium structure were performed with full use of symmetry in the respective symmetry coordinates. Since excitations within MRCI are subject to symmetry restrictions, the energies obtained at the same structure but using a different symmetry may differ slightly, thereby producing steps in the PES. In order to avoid artifacts introduced by these steps, it was necessary to calculate the whole PES within the same point group symmetry. Therefore, all cubic and quartic force fields were

calculated using C_s symmetry for all geometries. Only few energies had to be calculated in
C_1 symmetry, in order to obtain the $f_{4x4x5y5y}$ force constant. However, in this case MRCI
and ACPF are invariant with respect to the symmetry, meaning that energies obtained
using C_1 or C_s symmetry are equal within the energy convergence threshold of 10^{-10} E$_H$.
The large root mean square deviation of 43 cm^{-1} in the fit of Hochlaf *et al.*[34] might well
be the result of such steps in the PES.

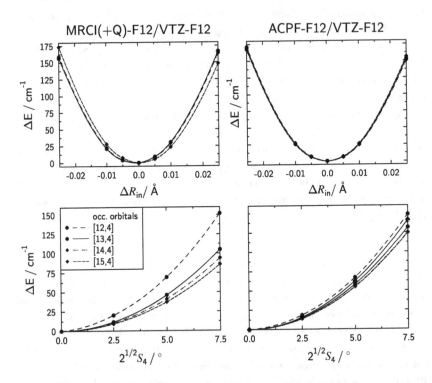

Figure 5: Cuts through the PES of C$_4$ applying active spaces of different size. Upper
panels were calculated using higher symmetry ($D_{\infty h}$) but occupied orbitals correspond to
those given in the key for C_s symmetry with the number of orbitals in the two irreducible
representations in the form [a', a'']. Solid line with club markers in panel D corresponds
to [13,4] active space with inclusion of ΔCV. Reference structure chosen: $R_{in} = 1.2920$ Å
and $R_{out} = 1.3121$ Å.

Figure 5 shows cuts through the PES along the $\Delta R_{in} = S_1$ and trans bending S_4 symmetry coordinates using active spaces of different size. The key gives the number of occupied orbitals in the irreducible representations $[a', a'']$ in C_s symmetry. In all calculations, seven orbitals in a' symmetry were treated as closed orbitals (occupation number is 2 in all determinants) and four of these are treated as core orbitals (which are not further optimized). The curves were calculated using a reference structure of $R_{in} = 1.2920$ Å and $R_{out} = 1.3121$ Å. MRCI(+Q)-F12/VTZ-F12 calculations (left) and ACPF-F12/VTZ-F12 (right) are compared and the solid line with filled circle markers denotes the active space used in this thesis. The plots for ΔR_{in} in the upper panel were calculated using Abelian D_{2h} symmetry but the occupied orbitals chosen correspond to those given in the key. Furthermore, these plots do not contain curves for the [12,4] active space because these calculations did not converge. When comparing the upper and lower panels, it is obvious that the R_{in} stretching coordinate is much less sensitive to the choice of the active space than the trans bending coordinate S_4, where it should be noted that the energy range is approximately the same for both panels. While the curves for the [13,4] and [14,4] occupied orbitals almost coincide, only slight differences are observed for the larger [15,4] variant in case of the MRCI(+Q) calculation and there is hardly a change at all for the curves as calculated by ACPF. The [15,4] calculations are around 35 times more expensive than the [13,4] calculations and are thus unmanageable with the available computational resources. Panel C clearly shows that the choice of [12,4] occupied orbitals is not suitable for the description of the bending coordinates. The difference amounts to almost 50 cm^{-1} for a displacement of 7.5 ° in the case of the MRCI(+Q) calculations as compared to the [13,4] active space but unfortunately even larger active spaces do not show convergence. It has to be noted that the behaviour of the curves without Davidson correction (not shown here) is comparable although the curves are steeper. In case of the ACPF calculations only a slight dependence on the size of the active space is observed with convergence almost reached for the [14,4] occupied orbitals active space. This may lead to the conclusion that the increased flexibility of the reference wavefunction is not fully exploited by the ACPF method. The inferiority of ACPF is further supported by the single-reference based calculations which are in good agreement with the MRCI(+Q) results as will be pointed out below. On the other hand there is no further theoretical support for the opposite conclusion that MRCI(+Q) overestimates the correlation. Panel D also displays the curve of a calculation where all electrons are correlated (solid line with club markers) using the [13,4] active space. In this case, the neglect of ΔCV is almost compensated by the choice of the active space that is slightly too small leading to a favourable error compensation at least for this coordinate.

Table 11: Basis set dependence of bond lengths (in Å), equilibrium rotational constants B_e (in MHz) and harmonic vibrational wavenumbers (in cm^{-1}) of C_4 as obtained by several single- and multi-reference methods.

method	R_{in}	R_{out}	B_e	$\omega_1(\sigma_g)$	$\omega_2(\sigma_g)$	$\omega_3(\sigma_u)$	$\omega_4(\pi_g)$	$\omega_5(\pi_u)$
MRCI-F12								
/VDZ-F12	1.28618	1.30601	4998.8	2146.4	948.7	1622.1	372.4	173.1
/VTZ-F12	1.28577	1.30534	5003.1	2144.5	949.6	1623.8	369.5	172.3
/VQZ-F12	1.28555	1.30507	5005.0	2145.5	950.2	1624.7	375.7	173.6
MRCI(+Q)-F12								
/VDZ-F12	1.29140	1.31194	4955.6	2111.5	927.8	1583.0	325.6	155.6
/VTZ-F12	1.29108	1.31131	4959.4	2109.2	928.7	1584.8	321.7	154.5
/VQZ-F12	1.29086	1.31106	4961.2	2110.2	929.1	1585.5	330.4	156.1
ACPF-F12								
/VDZ-F12	1.29237	1.31284	4948.5	2103.9	923.8	1576.8	340.5	169.3
/VTZ-F12	1.29203	1.31216	4952.6	2101.4	924.3	1578.3	358.4	168.9
/VQZ-F12	1.29176	1.31186	4954.8	2102.7	925.0	1579.7	366.8	170.5
CIPT2(+Q)								
/CVTZ	1.29151	1.31306	4950.1	2108.3	928.1	1580.2	336.8	161.9
/CVQZ	1.28887	1.30901	4976.3	2114.4	931.9	1589.4	331.2	157.7
/CV5Z	1.28820	1.30805	4983.1	2116.5	933.0	1592.4	330.4	156.4
ROHF-								
RCCSD(T)								
/VQZ	1.29360	1.31356	4941.5	2100.2	921.1	1930.9	375.7	171.1
/V5Z	1.29308	1.31267	4947.1	2100.8	921.2	1935.3	371.8	168.3
/V6Z	1.29292	1.31238	4952.4	2101.2	921.5	1935.6	372.4	168.6
UHF-								
UCCSD(T)								
/VTZ	1.29550	1.31672	4921.5	2097.5	916.1	1565.4	357.0	170.0
/VQZ	1.29300	1.31293	4946.2	2100.2	924.6	1571.5	351.1	164.5
/V5Z	1.29250	1.31204	4951.7	2100.8	924.7	1573.2	347.0	161.6

Having justified the [13,4] active space as the compromise between an accurate description and a computationally feasible choice, the basis set dependence of the structure and the harmonic wavenumbers as obtained by different methods will now be discussed (see Table 11). Although convergence is not yet reached, there is only little variance in the bond lengths for both MRCI-F12 and MRCI(+Q)-F12. The Davidson correction however has a large influence on both the harmonic vibrational wavenumbers and the rotational constant, where the former are higher by up to 45 cm^{-1} and the latter by 44 MHz when Davidson correction is not applied. Since the ΔCV contribution, which is omitted in these calculations, will further increase B_e as well as all harmonic vibrational wavenumbers, the values without Davidson correction are considered to be inferior. For these two methods most parameters considered in Table 11 follow clear trends with the exception of ω_1 and the bending vibrations. The reason for this was not resolved and is especially disturbing in the case of ω_4 where the difference between VTZ-F12 and VQZ-F12 amounts to about 9 cm^{-1} for the method with Davidson correction. With the exception of ω_4, the basis set dependence is not that pronounced which also holds true for ACPF-F12. The harmonic vibrational wavenumbers calculated with that method are somewhat lower in the case of the stretching wavenumbers and higher in the case of the bending wavenumbers when compared to MRCI(+Q)-F12. The trans bend wavenumber (ω_4) however increases with the basis set size and does not show convergence. All methods discussed so far suffer from the lack of core correlation which is included in the case of CIPT2(+Q). Since for these calculations all parameters follow certain trends with increasing basis set size, the oscillating behaviour discussed above might be due to approximations in the F12 formalism. The inclusion of core correlation increases all parameters in Table 11 as compared to the MRCI(+Q)-F12/VQZ-F12 method and is most pronounced for ω_1 and ω_3 (+6.4 cm^{-1} and +6.9 cm^{-1}, respectively). In the lower part of Table 11, results as obtained by the ROHF and UHF based single-reference methods are listed. While the results are almost identical for the symmetric stretching modes (and the structure, as already discussed), a different behaviour is observed for the two bending vibrations. Notably, ω_4 differs by more than 25 cm^{-1}. With the UHF-UCCSD(T) method, a decrease of the critical wavenumber ω_4 is observed with increasing basis set size, as opposed to the results obtained by ACPF-F12. In summary it can be concluded that a size-extensivity correction is necessary in order to arrive at reasonable wavenumbers and geometries in the case of C_4. While ACPF-F12 is in favourable agreement with both MRCI(+Q) and the single-reference methods in the case of the stretching wavenumbers, it seems to be less reliable for the description of the bending motion. The hybrid method CIPT2(+Q) appears to be a promising improvement over MRCI(+Q). In the case of single-reference methods, UHF-UCCSD(T) will be considered further, since it appears to be free from symmetry breaking problems and can be improved by inclusion of HC effects.

Table 12: Harmonic vibrational wavenumbers (in cm^{-1}) for C_4 as obtained by **Comp. 1.**

contribution[a]	$\omega_1(\sigma_g)$	$\omega_2(\sigma_g)$	$\omega_3(\sigma_u)$	$\omega_4(\pi_g)$	$\omega_5(\pi_u)$
fc-UCCSD(T)/VQZ	2100.30	924.63	1571.71	351.55	164.59
$+\Delta$CV/CVTZ	2107.82	928.39	1578.59	355.68	165.90
$+\Delta$DK/VTZ	2107.57	928.18	1578.07	355.45	165.85
$+\Delta$(Q)-(T)/VTZ	2084.77	919.02	1568.04	316.30	159.27
$+\Delta$Q-(Q)/VDZ (bend)[b]	2084.77	919.02	1568.04	323.49	162.07

[a]All CC calculations are based on UHF reference functions.
[b]The sum of all contributions is referred to as **Comp. 1.**

The effect of the smaller contributions on the harmonic vibrational wavenumbers is shown in Table 12 for a UHF based composite approach calculated in internal coordinates. While ΔCV increases all values, thereby being most important for ω_1 (+7.5 cm^{-1}) and ω_3 (+5.9 cm^{-1}), ΔDK is again of minor importance with a slightly decreasing effect on all wavenumbers. Inclusion of higher-order correlation (the sum of all effects being referred to as **Comp. 1** in the following) however leads to a significant decrease of all wavenumbers, especially in the case of ω_4 which is reduced by more than 30 cm^{-1} (almost 10 %). It has to be noted that the expensive ΔQ-(Q) calculations were only carried out for the quadratic term of the internal bending coordinate, where they are expected to have the largest influence. The inclusion of the smaller effects is further illustrated in Figure 6 where the major importance of ΔHC can be observed in the top panel for the θ dependence. It has to be noted that ΔCV is not yet converged with the CVTZ basis set. The effect on the two totally symmetric stretching coordinates is almost identical, where it has to be noted that in the middle panel both outer bonds are displaced simultaneously by the the same amount, leading to relative energies of about double the value of those in the lower panel.

The trans bend harmonic wavenumber has been found to be extremely method and basis set dependent. It is thus desirable to confirm the value of 323.49 cm^{-1} obtained by **Comp. 1** by another method that includes both ΔHC and ΔCV. Therefore, a composite calculation with ROHF based ae-UCCSD(T)/V5Z being the basis contribution and Δ(Q)-(T)/VTZ and ΔQ-(Q)/VDZ being subsequently added was carried out and resulted in $\omega_4 = 320.14$ cm^{-1}. This is a fair agreement given the different basis set sizes applied and speaks again in favor of a value for ω_4 which is at the lower end of the range discussed so far.

Spectroscopic parameters obtained from cubic force-fields are listed in Table 13. Given the large differences observed for the different methods in case of the harmonic vibrational wavenumbers, the agreement for the rotation-vibration interaction constants α_i is excellent especially in the case of α_{1-3}. In that regard, the large discrepancy between the value for α_1 as obtained by Hochlaf *et al.* and the value obtained in this thesis, which is about half of the former value, is remarkable. This can not only be explained by the different basis set size and ultimately leads to an opposite sign for ΔB_0. The value for $\alpha_3 = 21.3$ MHz as obtained by the **Comp. 1** method almost perfectly matches the experimental value of 21.13 MHz.[22] The rotation-vibration interaction constants for the bending normal coordinates appear to be more method dependent. Here, the CIPT2(+Q) values match the experimental counterparts best and lie in between those obtained by the ACPF-F12 methods and those obtained by MRCI(+Q) and **Comp. 1**. ΔB_0 thus only ranges between 10.0 MHz and 14.7 MHz. The average of that range together with the best estimate B_e value results in a B_0 value of 4980 MHz in perfect agreement with the experimental value. However it has to be noted that this agreement is build on extrapolations as well as averaging of different methods and is therefore somewhat fortuitous. The error range for B_0 can be conservatively estimated to be as high as 5 MHz. From the comparison of the values as obtained by ACPF-F12 with the VTZ and VQZ basis sets, it can be concluded that at least for the F12 methods the basis set dependence is small for these spectroscopic parameters and mostly due to the effect on the ω's. The quartic and sextic centrifugal distortion constant also show only little variance with the methods used. The l-type doubling constants q^e depend solely on the harmonic force field, so they can directly be related to the harmonic vibrational wavenumbers and are nicely reproduced by the MRCI(+Q)-F12 and the **Comp. 1** cubic force field. Since ω's for the

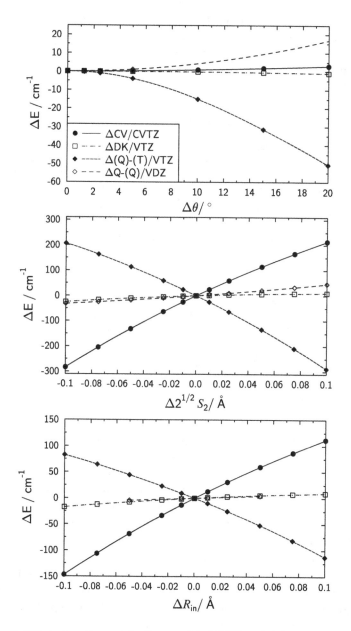

Figure 6: Effect of smaller contributions on $\Delta\theta$, $\Delta 2^{1/2} S_2$ and ΔR_{in} for C$_4$. All lines are spline smoothing functions with the exception of ΔQ-(Q) in the upper panel where the quadratic coefficient of the polynomial fit is shown.

Table 13: Spectroscopic parameters for C_4 obtained from cubic force fields.

	MRCI-F12 /VDZ-F12	MRCI(+Q)-F12 /VDZ-F12	ACPF-F12 /VTZ-F12	ACPF-F12 /VQZ-F12
α_1 / MHz	31.8	32.2	32.3	32.3
α_2 / MHz	8.4	8.7	8.8	8.8
α_3 / MHz	19.9	20.5	20.5	20.5
α_4 / MHz	-18.7	-21.7	-19.5	-19.1
α_5 / MHz	-21.6	-23.7	-22.0	-21.8
q_4^e / MHz	5.03	5.51	5.11	5.03
q_5^e / MHz	10.05	10.93	10.14	10.06
B_e / MHz	4998.70	4955.45	4952.53	4954.75
ΔB_0 / MHz	10.3	14.7	10.7	10.0
B_0 / MHz	5009.0	4970.1	4963.2	4974.8
D_e / kHz	616.7	628.2	631.5	631.6
H_e / mHz	0.032	0.029	0.028	0.028
	CIPT2(+Q) /CVTZ	Comp. 1 (see the text)	Hochlaf et al.[34] MRCI(+Q)/VTZ	experimental
α_1 / MHz	32.2	32.0	60	-
α_2 / MHz	8.7	8.7	21	-
α_3 / MHz	20.5	21.3	24	21.13[22]
α_4 / MHz	-20.2	-22.7	-18	-20.86[23]
α_5 / MHz	-22.5	-22.8	-18	-22.52[24]
q_4^e / MHz	5.37	5.55	5.55	5.52(19)[23]
q_5^e / MHz	10.53	10.52	10.07	10.96(12)[23]
B_e / MHz	4953.03	4949.22	4919.9	-
ΔB_0 / MHz	12.0	14.4	-15.3	-
B_0 / MHz	4965.0	4963.6	4904.6	4979.92(20)[24]
D_e / kHz	628.1	637.8	638	$D_0=855(40)$[24]
H_e / mHz	0.029	0.030	-	-

stretching vibrations as obtained by the composite method are considered to be reliable and an accurate equilibrium structure was obtained, the experimental q values may be used to derive semi-empirical harmonic vibrational wavenumbers for the bending modes. In order to carry out such a calculation, best estimate harmonic vibrational stretching frequencies were obtained from the UHF based composite method that was used for the determination of the equilibrium structure. In case of ω_3, the basis contribution had to be reduced to a CBS[3,4] extrapolation due to the lower symmetry of the S_3 coordinate. The resulting values are $\omega_1 = 2090.41$ cm^{-1}, $\omega_2 = 922.24$ cm^{-1} and $\omega_3 = 1574.48$ cm^{-1}. With these values and the experimental l-type doubling constants (see Table 13), the semi-empirical harmonic wavenumbers $\omega_4 = 330$ cm^{-1} and $\omega_5 = 159$ cm^{-1} were obtained. These values are in reasonable agreement with those calculated by **Comp. 1**. Although the formula for q^e used in this calculation (see eq. 34) is only approximative, the values obtained for the harmonic bending wavenumbers are in reasonable agreement with those calculated by **Comp. 1**. Unfortunately, due to the low symmetry of the respective coordinates, it was not yet possible to derive ω_4 and ω_5 at the same level of theory as for the

stretching vibrations. Thus, the values obtained by **Comp. 1** will be used as the best
estimate harmonic bending vibrational wavenumbers.

Table 14: Calculated anharmonic fundamental wavenumbers (in cm^{-1}) and anharmonic
contributions $\Delta_i = \nu_i - \omega_i$ (in cm^{-1}) for C_4.

cubic:[a]	MRCI(+Q)-F12	Comp. 1	Best estimate[b]
quartic:	MRCI(+Q)-F12	MRCI(+Q)-F12	MRCI(+Q)-F12
ν_1	2058.3	2033.1	2038.7
ν_2	940.2	932.8	935.6
ν_3	1557.8	1543.9	1550.3
ν_4	312.0	325.4	325.5
ν_5	152.3	164.9	165.2
Δ_1	-53.0	-51.7	-51.7
Δ_2	12.5	13.8	13.3
Δ_3	-25.2	-24.1	-24.2
Δ_4	-14.6	1.9	2.0
Δ_5	-3.5	2.9	3.1
cubic:[a]	ACPF-F12	Comp. 1	Best estimate[b]
quartic:	ACPF-F12	ACPF-F12	ACPF-F12
ν_1	2042.9	2025.8	2031.4
ν_2	937.4	930.2	933.0
ν_3	1547.8	1539.0	1545.3
ν_4	356.9	333.0	333.2
ν_5	167.0	163.4	163.7
Δ_1	-58.7	-59.0	-59.0
Δ_2	13.0	11.2	10.8
Δ_3	-30.5	-29.0	-29.2
Δ_4	-1.7	9.6	9.7
Δ_5	-2.0	1.4	1.6

[a]Here, *cubic* denotes the full cubic force field as well as
diagonal quartic force constants.
[b]Best estimate structure and harmonic wavenumbers (see the text)
are used along with **Comp. 1** cubic force-field

The anharmonic vibrational wavenumbers and corresponding anharmonic contributions were calculated using both the MRCI(+Q)-F12/VDZ-F12 and the ACPF-F12/VTZ-
F12 quartic force field. When the values as obtained by the two methods are compared
(see Table 14), the anharmonic contributions Δ_2 and Δ_5 differ only slightly, while the
differences are quite substantial for the remaining three vibrations. In case of the stretching modes ACPF-F12 gives values that are about 5 cm^{-1} lower than those obtained by
MRCI(+Q)-F12. For the trans bending vibration a large difference between these two
methods is observed, where the MRCI(+Q)-F12 values is 13 cm^{-1} lower. The replacement
of the force constants up to cubic and diagonal quartic ones (in internal coordinates) by
those obtained with the **Comp. 1** method, has only a minor effect on Δ_{1-3} but even
leads to a sign change in case of the bending frequencies. This is especially due to changes
in the bending anharmonic constants that result upon the replacement of these force con-

Table 15: Anharmonic constants (in cm^{-1}), quartic force constants in normal coordinates (in cm^{-1}) and quartic force constants in internal coordinates (in atomic units) for C_4 as calculated by different methods.

cubic:	ACPF-F12	Comp. 1	MRCI(+Q)-F12	Comp. 1
quartic:	ACPF-F12	ACPF-F12	MRCI(+Q)-F12	MRCI(+Q)-F12
χ_{44}	-0.9	2.9	-5.7	-2.0
χ_{45}	-1.2	1.4	-2.6	1.9
$\phi_{4444}/16$	14.0	19.9	10.5	15.0
$\phi_{4455}^{(+)}/4$	25.1	34.0	29.0	34.6
C_{4444}	0.009023	0.014143	0.004135	0.014143
C_{4445}		-0.003629		0.003690
C_{4455}		0.011150		0.003240
C_{4477}		-0.001190		-0.004091

stants. In their VPT2 study on the vibrational frequencies of acetylene, Martin et al.[80] found the diagonal anharmonic constant χ_{44} to be extremely basis set dependent and accounted the sensitivity almost exclusively to the quartic terms in the corresponding formula which depends on $\phi_{4444}/16$ (see eq. 19). Comparing χ_{44} and the off-diagonal bending anharmonic constant χ_{45} in Table 15 as obtained by the different methods shows that an analogous observation can be made for C_4. The largest part of the changes in the anharmonic constants depends on the changes in the diagonal quartic force constants in normal coordinates which holds true for both methods. The diagonal force constants in normal coordinates are in turn mostly dependent on the diagonal and off-diagonal quartic bending potential terms. Unfortunately, the large discrepancies between the potential terms in question cannot be resolved and there is no theoretical hint for preferring one over the other. The values given in Table 15 were checked by several reduced fits with varying choice of points but the discrepancies were maintained. It has to be noted however that the C_{4444} potential term obtained by Comp. 1 lacks the contribution of ΔQ-(Q) and is thus probably too high. These observations further confirm the pathologically difficult description of the bending coordinates for C_4, an issue that could not be resolved satisfyingly in this thesis. Further anharmonic constants as calculated by the "best estimate" methods can be found in the appendix.

The best estimate values of Table 14 are based on the equilibrium structure as obtained by the UHF based composite method (see Table 10), the Comp. 1 cubic and diagonal quartic force field and the best estimate harmonic wavenumbers given above. Using the off-diagonal quartic force-constants as obtained by MRCI(+Q)-F12 and ACPF-F12 results in the values given in the last column. Since again there is no sound argument to prefer one set of values over the other, the average of both in combination with a conservative error estimation is the most that can be achieved with the current data. Table 16 collects the best estimate theoretical values for experimentally observable parameters along with the most reliable experimental data.

Table 16: Best estimate theoretical values for experimentally observable parameters of C_4 along with experimental counterparts.

	theor.	exp.		theor.	exp.
ν_1 / cm^{-1}	2035.1 ± 10	2057 ± 50[a][18]	α_1 / MHz	32.0	
ν_2 / cm^{-1}	934.3 ± 5		α_2 / MHz	8.7	
ν_3 / cm^{-1}	1547.8 ± 10	$1548.6128(4)$[23]	α_3 / MHz	21.3	21.13[22]
ν_4 / cm^{-1}	329.4 ± 10	323 ± 50[a][18]	α_4 / MHz	-22.7	-20.86[23]
ν_5 / cm^{-1}	164.5 ± 5	172.4[b][20]	α_5 / MHz	-22.8	-22.52[24]
B_0 / MHz	4980 ± 5	$4979.92(20)$[23]	q_4 / MHz	5.60	5.52(19)[23]
			q_5 / MHz	10.60	10.96(12)[23]

[a]Values obtained from photoelectron spectroscopy.
[b]Argon-matrix value.

Only two experimental values in Table 16 are highly accurate, namely B_0 and ν_3 and both values are in excellent agreement with the best estimate theoretical values. However, this agreement is somewhat fortuitous given the large error bars and the approximations made in the derivation of these parameters. It has to be emphasized that neither the calculations nor the analysis (e.g. the construction of the composite methods) was biased to fit these experimental values but a sound overall description of the PES was pursued. The challenging electronic structure and the strong method dependence of the trans bending motion in particular did unfortunately not fully allow to achieve that aim. However, the results presented are derived from state-of-the-art electronic structure calculations and both these calculations and the further analysis is superior to all theoretical studies published to date. The good agreement with the existing experimental data raises confidence that further experimental studies will confirm the predicted parameters to lie within the (rather large) error bars. The already criticised estimation of $\nu_4 = 352 \pm 15$ cm^{-1} by Moazzen-Ahmadi and Thong[23] has shown to be too optimistic with regard to the error bars and a lower value more in the range of that obtained from photoelectron spectroscopy[18] is more likely. Unfortunately, the experimental observation of that IR inactive vibration is unlikely in near future. Concerning the astronomical observation of the IR band at 174 cm^{-1}, the values obtained in this thesis do not support the assignment to C_4. This is in agreement with the earlier criticism based on the large spin-orbit interaction constant required to simulate the band-shape observed.[17,34]

5 Conclusion and outlook

The aim of this thesis was to derive highly accurate spectroscopic parameters for the interstellar molecules l-C_3H^+ and C_4 from PESs that were calculated using state-of-the-art electronic structure calculations. While this aim was certainly achieved in the case of l-C_3H^+, the linear carbon chain C_4 in its $^3\Sigma_g^-$ ground state proved to be more challenging. However, also in this case it was possible to calculate spectroscopic parameters that agree well with the available experimental data especially in the case of the only two highly accurate values, namely B_0 and ν_3. The pathologically difficult description of the bending coordinates and the resulting wavenumbers was solved to some extent and values of $\nu_4 = 329.4 \pm 10$ cm^{-1} and $\nu_5 = 164.5 \pm 5$ cm^{-1} were predicted. The latter value questions the assignment of an IR band at 174 cm^{-1} detected in interstellar and circumstellar clouds to C_4.[17] Future work should focus on the calculation of the spin-orbit interaction constant required to fit the experimentally observed data and search for other likely candidates among the linear carbon chains. Since the spectrum of methods available to date does not allow for a description of the electronic structure as accurate as for closed-shell molecules, more advanced multi-reference methods are required. The internally contracted multireference coupled-cluster method as derived and implemented by Köhn and Hanauer might be the most promising candidate in that respect.[103,104] Despite the fact that it was not possible to reach the same level of accuracy compared to l-C_3H^+ and other closed-shell molecules,[78,79,105] valuable predictions could still be made. To the best of my knowledge there are only few highly accurate theoretical studies on the spectroscopic parameters of open-shell molecules that are based on UHF-coupled-cluster approaches and also include the effects of higher-order correlation. The results of this study might then also serve as a reference for further studies on the rovibrational spectroscopy of molecules with open-shell ground states.

In the case of l-C_3H^+, it was shown that the inclusion of HC effects is mandatory in order to obtain reliable spectroscopic parameters even for closed-shell molecules. It is apparent from the results as obtained by the single- and multi-reference approaches that the treatment of higher-order excitations within the CC hierarchy is capable of fully compensate for the deficiencies of the reference wavefunction when the multi-configurational character is not too high. Since the results for l-C_3H^+ are in such excellent agreement to the experimental values and very consistent between the different approaches, it is desirable to also use the calculated PESs in variational calculations of the rovibrational energies and wavefunctions. This will further enlighten the weak Fermi resonance between ν_1 and $\nu_2 + \nu_3$ and enable to obtain precise values for the bending anharmonic wavenumbers and intensities, as well. It will furthermore be interesting to see if the low-lying CCC bending vibration ν_5 is well described by VPT2. The five-dimensional potentials calculated in this thesis can easily be extended to full six-dimensional potentials by addition of respective terms for the torsion coordinate. Corresponding calculations and fits are underway and the variational calculations will follow in due course.

The VPT2 program **4Lin** that was programmed in the course of this thesis reliably calculates the spectroscopic parameters and the possibility to quickly exchange parts of the potential in both internal and normal coordinate space proved to be very helpful. VPT2 theory has again proven to be a valuable tool for the calculation of spectroscopic parameters in the case of semirigid molecules. The own implementation of the VPT2 formulas further helped to identify problematic terms in the PESs, a task that would not have been as easy with a black-box program.

6 References

[1] P. Botschwina, M. Horn, J. Flügge, S. Seeger, *J. Chem. Soc. Faraday Trans.* **1993**, *89*, 2219–2230.

[2] D. K. Bohme in *Structure/Reactivity and Thermochemistry of Ions*, (Eds.: P. Ausloos, S. G. Lias), Reidel, Dordrecht, **1987**.

[3] D. K. Bohme in *Rate Coefficients in Astrochemistry*, (Eds.: T. J. Millar, D. A. Williams), Kluwer, Dordrecht, **1988**.

[4] B. E. Turner, E. Herbst, R. Terzieva, *The Astrophysical Journal Supplement Series* **2000**, *126*, 427.

[5] V. Wakelam, I. Smith, E. Herbst, J. Troe, W. Geppert, H. Linnartz, K. Öberg, E. Roueff, M. Agúndez, P. Pernot, H. Cuppen, J. Loison, D. Talbi, English, *Space Science Reviews* **2010**, *156*, 13–72.

[6] Pety, J., Gratier, P., Guzmán, V., Roueff, E., Gerin, M., Goicoechea, J. R., Bardeau, S., Sievers, A., Le Petit, F., Le Bourlot, J., Belloche, A., Talbi, D., *A&A* **2012**, *548*, A68.

[7] S. Ikuta, *The Journal of Chemical Physics* **1997**, *106*, 4536–4542.

[8] B. A. McGuire, P. B. Carroll, R. A. Loomis, G. A. Blake, J. M. Hollis, F. J. Lovas, P. R. Jewell, A. J. Remijan, *The Astrophysical Journal* **2013**, *774*, 56.

[9] X. Huang, R. C. Fortenberry, T. J. Lee, *The Astrophysical Journal Letters* **2013**, *768*, L25.

[10] R. C. Fortenberry, X. Huang, T. D. Crawford, T. J. Lee, *The Astrophysical Journal* **2013**, *772*, 39.

[11] B. A. McGuire, P. B. Carroll, P. Gratier, V. Guzmán, J. Pety, E. Roueff, M. Gerin, G. A. Blake, A. J. Remijan, *The Astrophysical Journal* **2014**, *783*, 36.

[12] S. Brünken, L. Kluge, A. Stoffels, O. Asvany, S. Schlemmer, *The Astrophysical Journal Letters* **2014**, *783*, L4.

[13] P. Botschwina, C. Stein, P. Sebald, B. Schröder, R. Oswald, *The Astrophysical Journal* **2014**, *787*, 72.

[14] T. F. Giesen, A. O. V. Orden, J. D. Cruzan, R. A. Provencal, R. J. Saykally, R. Gendriesch, F. Lewen, G. Winnewisser, *The Astrophysical Journal Letters* **2001**, *551*, L181.

[15] J. P. Maier, N. M. Lakin, G. A. H. Walker, D. A. Bohlender, *The Astrophysical Journal* **2001**, *553*, 267.

[16] P. F. Bernath, K. H. Hinkle, J. J. Keady, *Science* **1989**, *244*, 562–564.

[17] J. Cernicharo, J. R. Goicoechea, Y. Benilan, *The Astrophysical Journal Letters* **2002**, *580*, L157.

[18] C. Xu, G. R. Burton, T. R. Taylor, D. M. Neumark, *The Journal of Chemical Physics* **1997**, *107*, 3428–3436.

[19] L. N. Shen, W. R. M. Graham, *The Journal of Chemical Physics* **1989**, *91*, 5115–5116.

[20] P. A. Withey, L. N. Shen, W. R. M. Graham, *The Journal of Chemical Physics* **1991**, *95*, 820–823.

[21] J. R. Heath, R. J. Saykally, *The Journal of Chemical Physics* **1991**, *94*, 3271–3273.

[22] N. Moazzen-Ahmadi, J. J. Thong, A. R. W. McKellar, *The Journal of Chemical Physics* **1994**, *100*, 4033–4038.

[23] N. Moazzen-Ahmadi, J. Thong, *Chemical Physics Letters* **1995**, *233*, 471–476.

[24] S. Gakwaya, Z. Abusara, N. Moazzen-Ahmadi, *Chemical Physics Letters* **2004**, *398*, 564–571.

[25] M. Algranati, H. Feldman, D. Kella, E. Malkin, E. Miklazky, R. Naaman, Z. Vager, J. Zajfman, *The Journal of Chemical Physics* **1989**, *90*, 4617–4618.

[26] D. Zajfman, H. Feldman, O. Heber, D. Kella, D. Majer, Z. Vager, R. Naaman, *Science* **1992**, *258*, 1129–1131.

[27] K. S. Pitzer, E. Clementi, *Journal of the American Chemical Society* **1959**, *81*, 4477–4485.

[28] E. Clementi, *Journal of the American Chemical Society* **1961**, *83*, 4501–4505.

[29] D. E. Bernholdt, D. H. Magers, R. J. Bartlett, *The Journal of Chemical Physics* **1988**, *89*, 3612–3617.

[30] J. M. L. Martin, T. J. Lee, P. R. Taylor, J. François, *The Journal of Chemical Physics* **1995**, *103*, 2589–2602.

[31] P. Botschwina, *Journal of Molecular Spectroscopy* **1997**, *186*, 203–204.

[32] P. Botschwina, *Chemical Physics Letters* **2006**, *421*, 488–493.

[33] H. Massó, M. L. Senent, P. Rosmus, M. Hochlaf, *The Journal of Chemical Physics* **2006**, *124*, 234304.

[34] M. L. Senent, H. Massó, M. Hochlaf, *The Astrophysical Journal* **2007**, *670*, 1510.

[35] H.-J. Werner, P. J. Knowles, R. Lindh, R. F. Manby, M. Schütz, et al., MOLPRO version 2012.1 a package of ab initio programs, see http://www.molpro.net (accessed Mar 24 2014), **2012**.

[36] F. Jensen, *Introduction to Computational Chemistry*, 2nd ed., John Wiley & Sons, Chichester, **2007**.

[37] J. Čížek, *The Journal of Chemical Physics* **1966**, *45*, 4256–4266.

[38] H. Kümmel, English, *Theoretica chimica acta* **1991**, *80*, 81–89.

[39] R. J. Bartlett, *The Journal of Physical Chemistry* **1989**, *93*, 1697–1708.

[40] K. Raghavachari, G. W. Trucks, J. A. Pople, M. Head-Gordon, *Chemical Physics Letters* **1989**, *157*, 479–483.

[41] P. J. Knowles, C. Hampel, H.-J. Werner, *The Journal of Chemical Physics* **1993**, *99*, 5219–5227.

[42] P. J. Knowles, C. Hampel, H.-J. Werner, *The Journal of Chemical Physics* **2000**, *112*, 3106–3107.

[43] MRCC, a string based quantum chemical program suite written by M. Kállay, see http://www.mrcc.hu (accessed Mar 24 2014).

[44] E. Hylleraas, German, *Zeitschrift für Physik* **1929**, *54*, 347–366.

[45] W. Kutzelnigg, W. Klopper, *The Journal of Chemical Physics* **1991**, *94*, 1985–2001.

[46] V. Termath, W. Klopper, W. Kutzelnigg, *The Journal of Chemical Physics* **1991**, *94*, 2002–2019.

[47] W. Klopper, W. Kutzelnigg, *The Journal of Chemical Physics* **1991**, *94*, 2020–2030.

[48] S. Ten-no, *Chemical Physics Letters* **2004**, *398*, 56–61.

[49] T. B. Adler, G. Knizia, H.-J. Werner, *The Journal of Chemical Physics* **2007**, *127*, 221106.

[50] G. Knizia, T. B. Adler, H.-J. Werner, *The Journal of Chemical Physics* **2009**, *130*, 054104.

[51] H.-J. Werner, W. Meyer, *The Journal of Chemical Physics* **1980**, *73*, 2342–2356.

[52] H.-J. Werner, P. J. Knowles, *The Journal of Chemical Physics* **1985**, *82*, 5053–5063.

[53] P. J. Knowles, H.-J. Werner, *Chemical Physics Letters* **1985**, *115*, 259–267.

[54] W. Meyer in *Modern Theoretical Chemistry*, (Ed.: H. F. Schaefer III), Plenum, New York, **1977**, p. 413.

[55] H.-J. Werner, E. Reinsch, *The Journal of Chemical Physics* **1982**, *76*, 3144–3156.

[56] P. J. Knowles, H.-J. Werner, *Chemical Physics Letters* **1988**, *145*, 514–522.

[57] H.-J. Werner, P. J. Knowles, *The Journal of Chemical Physics* **1988**, *89*, 5803–5814.

[58] E. R. Davidson, D. W. Silver, *Chemical Physics Letters* **1977**, *52*, 403–406.

[59] R. J. Gdanitz, R. Ahlrichs, *Chemical Physics Letters* **1988**, *143*, 413–420.

[60] T. Shiozaki, G. Knizia, H.-J. Werner, *The Journal of Chemical Physics* **2011**, *134*, 034113.

[61] P. Celani, H. Stoll, H.-J. Werner, P. Knowles, *Molecular Physics* **2004**, *102*, 2369–2379.

[62] T. H. Dunning, *The Journal of Chemical Physics* **1989**, *90*, 1007–1023.

[63] K. A. Peterson, T. B. Adler, H.-J. Werner, *The Journal of Chemical Physics* **2008**, *128*, 084102.

[64] K. E. Yousaf, K. A. Peterson, *The Journal of Chemical Physics* **2008**, *129*, 184108.

[65] F. Weigend, A. Köhn, C. Hättig, *The Journal of Chemical Physics* **2002**, *116*, 3175–3183.

[66] F. Weigend, *Phys. Chem. Chem. Phys.* **2002**, *4*, 4285–4291.

[67] M. Douglas, N. M. Kroll, *Annals of Physics* **1974**, *82*, 89–155.

[68] B. A. Hess, *Phys. Rev. A* **1986**, *33*, 3742–3748.

[69] G. Jansen, B. A. Hess, *Phys. Rev. A* **1989**, *39*, 6016–6017.

[70] W. A de Jong, R. J. Harrison, D. A. Dixon, *The Journal of Chemical Physics* **2001**, *114*, 48–53.

[71] D. E. Woon, T. H. Dunning, *The Journal of Chemical Physics* **1995**, *103*, 4572–4585.

[72] D. Feller, J. A. Sordo, *The Journal of Chemical Physics* **2000**, *112*, 5604–5610.

[73] Y. J. Bomble, J. F. Stanton, M. Kállay, J. Gauss, *The Journal of Chemical Physics* **2005**, *123*, 054101.

[74] M. Kállay, P. R. Surján, *The Journal of Chemical Physics* **2001**, *115*, 2945–2954.

[75] J. Gauss, A. Tajti, M. Kállay, J. F. Stanton, P. G. Szalay, *The Journal of Chemical Physics* **2006**, *125*, 144111.

[76] CFOUR, a quantum chemical program package written by J.F. Stanton, J. Gauss, M.E. Harding, P.G. Szalay with contributions from A.A. Auer, R.J. Bartlett, U. Benedikt, C. Berger, D.E. Bernholdt, Y.J. Bomble, L. Cheng, O. Christiansen, M. Heckert, O. Heun, C. Huber, T.-C. Jagau, D. Jonsson, J. Jusélius, K. Klein, W.J. Lauderdale, D.A. Matthews, T. Metzroth, L.A. Mück, D.P. O'Neill, D.R. Price, E. Prochnow, C. Puzzarini, K. Ruud, F. Schiffmann, W. Schwalbach, C. Simmons, S. Stopkowicz, A. Tajti, J. Vázquez, F. Wang, J.D. Watts and the integral packages MOLECULE (J. Almlöf and P.R. Taylor), PROPS (P.R. Taylor), ABACUS (T. Helgaker, H.J. Aa. Jensen, P. Jørgensen, and J. Olsen), and ECP routines by A. V. Mitin and C. van Wüllen. For the current version see http://www.cfour.de.

[77] J. K. Watson, *Molecular Physics* **1970**, *19*, 465–487.

[78] P. Sebald, A. Bargholz, R. Oswald, C. Stein, P. Botschwina, *The Journal of Physical Chemistry A* **2013**, *117*, 9695–9703.

[79] P. Sebald, C. Stein, R. Oswald, P. Botschwina, *The Journal of Physical Chemistry A* **2013**, *117*, 13806–13814.

[80] J. M. L. Martin, T. J. Lee, P. R. Taylor, *The Journal of Chemical Physics* **1998**, *108*, 676–691.

[81] R. C. Herman, W. H. Shaffer, *The Journal of Chemical Physics* **1948**, *16*, 453–465.

[82] H. H. Nielsen, *Rev. Mod. Phys.* **1951**, *23*, 90–136.

[83] S. Califano, *Vibrational States*, John Wiley & Sons, London, **1976**.

[84] W. D. Allen, Y. Yamaguchi, A. G. Császár, D. A. Clabo Jr., R. B. Remington, H. F. Schaefer III, *Chemical Physics* **1990**, *145*, 427–466.

[85] H. H. Nielsen, *Phys. Rev.* **1941**, *60*, 794–810.

[86] J. F. Gaw, A. Willets, W. H. Green, N. C. Handy in *Advances in Molecular Vibrations and Collision Dynamics*, (Ed.: J. M. Bowman), JAI Press, Inc., Greenwich, **1991**, pp. 169–187.

[87] V. Barone, M. Biczysko, J. Bloino, *Phys. Chem. Chem. Phys.* **2014**, *16*, 1759–1787.

[88] G. Strey, I. Mills, *Journal of Molecular Spectroscopy* **1976**, *59*, 103–115.

[89] P. Botschwina, *Journal of Molecular Structure: THEOCHEM* **2005**, *724*, 95–98.

[90] P. Botschwina, R. Oswald, *The Journal of Chemical Physics* **2008**, *129*, 044305.

[91] P. Botschwina, R. Oswald, *Journal of Molecular Spectroscopy* **2009**, *254*, 47–52.

[92] P. Thaddeus, C. A. Gottlieb, H. Gupta, S. Brünken, M. C. McCarthy, M. Agúndez, M. Guélin, J. Cernicharo, *The Astrophysical Journal* **2008**, *677*, 1132.

[93] T. B. Tang, H. Inokuchi, S. Saito, C. Yamada, E. Hirota, *Chemical Physics Letters* **1985**, *116*, 83–85.

[94] J. Krieg, V. Lutter, C. P. Endres, I. H. Keppeler, P. Jensen, M. E. Harding, J. Vázquez, S. Schlemmer, T. F. Giesen, S. Thorwirth, *The Journal of Physical Chemistry A* **2013**, *117*, 3332–3339.

[95] P. Sebald, PhD thesis, Universität Kaiserslautern, **1990**.

[96] P. Botschwina, *Chemical Physics* **1983**, *81*, 73–85.

[97] P. Botschwina, *Chemical Physics Letters* **1984**, *107*, 535–541.

[98] P. Botschwina, R. Oswald, G. Rauhut, *Phys. Chem. Chem. Phys.* **2011**, *13*, 7921–7929.

[99]   O. Asvany, S. Brünken, L. Kluge, S. Schlemmer, English, *Applied Physics B* **2014**, *114*, 203–211.

[100] M. Kállay, J. Gauss, *The Journal of Chemical Physics* **2008**, *129*, 144101.

[101] J. M. Martin, *Chemical Physics Letters* **1996**, *259*, 669–678.

[102] D. Feller, K. A. Peterson, J. Grant Hill, *The Journal of Chemical Physics* **2011**, *135*, 044102.

[103] M. Hanauer, A. Köhn, *The Journal of Chemical Physics* **2011**, *134*, 204111.

[104] W. Liu, M. Hanauer, A. Köhn, *Chemical Physics Letters* **2013**, *565*, 122–127.

[105] P. Botschwina, P. Sebald, B. Schröder, A. Bargholz, K. Kawaguchi, T. Amano, *Journal of Molecular Spectroscopy* **2014**, *302*, 3 –8.

[106] A. Hoy, I. Mills, G. Strey, *Molecular Physics* **1972**, *24*, 1265–1290.

[107] E. B. Wilson, J. C. Decius, P. C. Cross, *Molecular Vibrations*, Dover, Inc., New York, **1980**.

[108] Intel, Math Kernel Library, http://developer.intel.com/software/products/mkl/.

[109] J. K. G. Watson in *Vibrational Spectra and Structure, Vol. 6*, (Ed.: J. R. Durig), Elsevier, Amsterdam, **1977**, Chapter 1.

[110] L. Hedberg, I. Mills, *Journal of Molecular Spectroscopy* **1993**, *160*, 117–142.

[111] V. Barone, *The Journal of Chemical Physics* **2005**, *122*, 014108.

A Appendix

A.1 Description of 4Lin: a VPT2 program for linear molecules with up to four atoms

The program 4Lin enables to deduce various rovibrational spectroscopic constants of linear molecules from a given anharmonic force field. So far it has been tested and runs stable for linear molecules with up to four atoms. The number of atoms, their masses (in atomic mass units u), and the bond lengths (in Å) have to be specified in the input. The ordering of the bond lengths corresponds to the ordering of the masses. The force constants are extracted from potential terms of an expansion of the form

$$V = \sum_{ijklm} C_{ijklm} \Delta r_1^i \Delta r_2^j \Delta r_3^k \Delta(\sin\theta_1)^l \Delta(\sin\theta_2)^m \qquad l, m : even \tag{49}$$

The Δr_i and $\Delta\theta_j$ are internal displacement coordinates for the bond stretching and angle bending, respectively. In case of the bending coordinates it is necessary to choose $\sin\theta$ as a coordinate since only this choice ensures the correct symmetry.[80,84,106] However, this choice affects only two of the quartic bending force constants in case of four-atomic molecules. Potential terms have to be entered in atomic units. The program then calculates the force constants from the potential terms so that the potential can be written as:

$$V = \sum_{ij} 1/2 f_{ij} s_i s_j + \sum_{ijk} 1/6 f_{ijk} s_i s_j s_k + \sum_{ijkl} 1/24 f_{ijkl} s_i s_j s_k s_l... \tag{50}$$

where s_i denotes any of the two types of coordinates described above. The 4Lin input requires the terms for degenerate (bending) coordinates only once, thereby reducing the number of coordinates to 3 in case of 3-atomic molecules and 5 in case of 4-atomic molecules. While the degenerate force constants can be copied in case of 3-atomic molecules to obtain the missing coordinate, certain symmetry relations must be considered to completely define the quartic force-field in case of 4-atomic molecules:[88]

$$f_{4x4x4x4x} = f_{4y4y4y4y} = 3f_{4x4x4y4y} \tag{51}$$

$$f_{4x4x4x5x} = 3f_{4x4x4y5y} \qquad etc. \tag{52}$$

$$f_{4x4y5x5y} = 1/2 \cdot (f_{4x4x5x5x} - f_{4x4x5y5y}) \tag{53}$$

In the above equations, only $f_{4x4x5y5y}$ cannot be obtained from a five-dimensional potential and has to be entered separately.

Having extracted the quadratic part of the force field (the F-matrix), the program solves the harmonic problem using the method described in Califano's book.[83]

If $\mathbf{s}$ denotes a vector containing the internal coordinates and $\mathbf{x}$ is the vector of the Cartesian coordinates of a given molecule, a linear relation between these coordinates can be found if we restrict our analysis to infinitesimal displacements

$$\mathbf{s} = \mathbf{Bx} \qquad . \tag{54}$$

The elements of the $\mathbf{B}$-matrix depend on the types of coordinates which are restricted to bond stretching coordinates and linear angle bending coordinates in the case of linear molecules. Formulas for the calculations of the $\mathbf{B}$-matrix elements can be found in Califano's book[83] or the original treatment by Wilson, Decius and Cross.[107] The next step is the calculation of the $\mathbf{G}$-matrix which is defined by

$$\mathbf{G} = \mathbf{BM}^{-1}\mathbf{B}^T \tag{55}$$

where $\mathbf{M}^{-1}$ is a diagonal matrix of the reciprocal atomic masses. The secular equation for the vibrational problem in internal coordinates can then be written as[83]

$$\mathbf{GFL} = \mathbf{L}\boldsymbol{\Lambda} \tag{56}$$

where $\mathbf{L}$ is the transformation from internal coordinates to normal coordinates $\mathbf{Q}$ ($\mathbf{s} = \mathbf{LQ}$), and $\boldsymbol{\Lambda}$ is a diagonal matrix of the eigenvalues of the harmonic problem, that is the harmonic force constants. Since the $\mathbf{GF}$ product matrix is not symmetric, it is useful to first bring the secular equation to a symmetric form. The harmonic problem is solved such that the harmonic vibrational wavenumbers are determined and the normal coordinates are defined.[4]
4Lin now sorts the harmonic vibrational wavenumbers and the L-matrix to spectroscopic order in the case of four atomic molecules, thereby discriminating between molecules with $D_{\infty h}$ or $C_{\infty v}$ symmetry.

The Coriolis-coupling constants ζ_{ij}^{σ} are defined by the harmonic force field alone and can thus be calculated at this point via

$$\zeta^{\sigma} = \mathbf{L}^{-1}\mathbf{D}M^{\sigma}\mathbf{D}^T(\mathbf{L}^{-1})^T \tag{57}$$

$$\mathbf{D} = \mathbf{BM}^{-1/2} \tag{58}$$

where the M^{σ} matrix is of dimension $3N \times 3N$ and consists of N identical 3×3 matrices defined in Califano, p.97.[83]
The equilibrium rotational constant can already be calculated from the geometry alone

---

[4]The routines for solving the eigenvalue problem (dsyev) and matrix inversion (dgetri) were taken from the Intel MKL library.[108]

since (in atomic units) it is defined as:

$$B_e = \frac{1}{2I} \tag{59}$$

I denoting the moment of inertia. The centrifugal distortion constant is calculated following the method first described by Watson[109] which can be greatly simplified in the case of linear molecules (see als eq. 30).[110]

The next crucial step is to transform the anharmonic force field to normal coordinates. This is a non-linear transformation and Hoy, Mills and Strey have derived formulas by differentiating the defining formulas for the different types of coordinates with respect to the normal coordinates.[106] This leads to a L-tensor formalism, where all higher-order elements of the L-tensor are derived from the first order elements which is the familiar L-matrix. The most convenient way to formulate the equations defining the L-tensor elements is via an additional transformation matrix $\mathbf{P}$ whose elements are defined as:

$$P_{\alpha i}^r = (l_{\alpha b}^r / m_b^{1/2}) - (l_{\alpha a}^r / m_a^{1/2}) \tag{60}$$

where the $l_{\alpha a/b}^r$ are elements of the transformation matrix from normal coordinates to mass-weigthed Cartesian displacement coordinates defined by

$$\mathbf{l} = \mathbf{M}^{-1/2} \mathbf{B}^T (\mathbf{L}^{-1})^T \qquad . \tag{61}$$

The indices a, b in eq. 60 label atoms, i labels the bonds from left to right as specified in the input and r denotes the normal coordinate.

In this formalism, the formulas for the L-tensor elements for the bond stretching coordinates read:

$$L_i^r = \mathbf{r}_i \cdot \mathbf{P}_i^r / r_i \tag{62}$$

$$L_i^{rs} = (\mathbf{P}_i^r \cdot \mathbf{P}_i^s - L_i^r L_i^s) / r_i \tag{63}$$

$$L_i^{rst} = -(L_i^r L_i^{st} + L_i^s L_i^{rt} + L_i^t L_i^{rs}) / r_i \tag{64}$$

In case of the linear angle bending coordinates they take the slightly more complex form:

$$L_{ijA}^r = \mathbf{e}_A \cdot (\mathbf{P}_i^r \times \mathbf{r}_j + \mathbf{r}_i \times \mathbf{P}_j^r) / r_i r_j \tag{65}$$

$$L_{ijA}^{rs} = -L_{ijA}^r (L_i^s / r_i + L_j^s / r_j) - L_{ijA}^s (L_i^r / r_i + L_j^r / r_j) + \mathbf{e}_A \cdot (\mathbf{P}_i^r \times \mathbf{P}_j^s + \mathbf{P}_i^s \times \mathbf{P}_j^r) / r_i r_j \tag{66}$$

$$L_{ijA}^{rst} = - \sum_{\text{perm } r,s,t} \{ (L_i^r / r_i + L_j^r / r_j) L_{ijA}^{st} + L_{ijA}^r [(L_i^s L_j^t + L_i^t L_j^s) / r_i r_j + L_i^{st} / r_i + L_j^{st} / r_j] \} \tag{67}$$

Thus, the L-tensor depends solely on the quadratic force field because this defines the L-Matrix. The equations for the first order L-tensor elements can be used to check the correctness of the P-matrix.

The anharmonic force constants in normal coordinates are then obtained by summations over L-tensor elements multiplied by the respective anharmonic force constants in internal displacement coordinates.

$$\Phi^{rs} = \sum_{ij} f_{ij} L_i^r L_j^s \tag{68}$$

$$\Phi^{rst} = \sum_{ijk} f_{ijk} L_i^r L_j^s L_k^t + \sum_{ij} f_{ij}(L_i^{rs} L_j^t + L_i^{rt} L_j^s + L_i^{st} L_j^r) \tag{69}$$

$$\Phi^{rstu} = \sum_{ijkl} f_{ijkl} L_i^r L_j^s L_k^t L_l^u + \sum_{ijk} f_{ijk}(L_i^{rs} L_j^t L_k^u + L_i^{rt} L_j^s L_k^u + L_i^{ru} L_j^s L_k^t + L_i^{st} L_j^r L_k^u$$

$$+ L_i^{su} L_j^r L_k^t + L_i^{tu} L_j^r L_k^s) + \sum_{ij} f_{ij}(L_i^{rs} L_j^{tu} + L_i^{rt} L_j^{su} + L_i^{ru} L_j^{st}) \tag{70}$$

$$+ \sum_{ij} f_{ij}(L_i^{rst} L_j^u + L_i^{rsu} L_j^t + L_i^{rtu} L_j^s + L_i^{stu} L_j^r)$$

These force constants in normal coordinates Q_r are then transformed to dimensionless normal coordinates q_r to be appropriate for use in VPT2 formulas. The conversion between these two types of coordinates is given by

$$q_r = \sqrt{2\pi c \omega_r / \hbar} Q_r \tag{71}$$

so that conversion of the anharmonic force constants can be achieved through dividing by the square root of the respective harmonic force constants. 4Lin now calculates the spectroscopic parameters from the anharmonic constants via the formulas described in section 2.2.

4Lin allows for a composite approach in calculating the force constants. If any quartic force field is given as the potential, quadratic or cubic force constants in normal coordinate space may be replaced by ones that were obtained by another (maybe better) method or experimentally derived force constants. Furthermore, since the transformation depends only on the L-matrix elements, these may be replaced as well, leading to slightly different quartic force constants. The result is a composite approach in normal coordinate space that enables different electronic structure methods to define different orders of the potential expansion in normal coordinates.

4Lin is also able to detect Fermi type 1 ($2\omega_a \approx \omega_b$) and Fermi type 2 ($\omega_a + \omega_b \approx \omega_c$) resonances that lead to unphysically large anharmonicity constants. Following Barone[111] and Martin et al.[30] the program does not use the empirical criterium ($\Delta = 2\omega_a - \omega_b$ or $\Delta = \omega_a + \omega_b - \omega_c$) for the detection of a perturbation, but takes into account that the error in the perturbative treatment depends also on the numerator. The resulting formulas are

$$\Delta_{ab}^{\text{Fermi1}} = \frac{\phi_{aab}^4}{256(2\omega_a - \omega_b)^3} \tag{72}$$

for Fermi type 1 resonances and

$$\Delta_{abc}^{\text{Fermi2}} = \frac{\phi_{abc}^4}{64(\omega_a + \omega_b - \omega_c)^3} \tag{73}$$

for Fermi type 2 resonances. The default value for Δ is 10 cm^{-1}. If a resonance is detected, the formulas for the anharmonicity constants are then factored differently and the resonating terms are omitted allowing for a first-order treatment as described in section 2.2.[84,86,111]

The program has been extensively tested on published force fields for both 3- and 4-atomic linear molecules.[9,80,84] Future extensions might include the use of symmetry coordinates, detection of Coriolis resonances, allowing for the calculation of several isotopologues in one input, calculation of intensities and most importantly the extension to larger molecules.

A.2   Force fields for l-C_3H^+

The force fields as already transformed to the equilibrium geometry will be given as potential terms in atomic units in case of the multi-reference methods. In case of the single-reference composite methods, the untransformed potential terms will be quoted allowing to check the behaviour upon neglect of some of the contributions. See eq. 38 and 44 for a definition of the PEF terms.

Table 17: Potential terms up to quartic ones (in atomic units) for l-C_3H^+ as obtained by MRCI(+Q)-F12/VQZ-F12.[a]

i	j	k	l	m	C_{ijklm}	i	j	k	l	m	C_{ijklm}
2	0	0	0	0	0.2676369	1	1	2	0	0	0.0038193
3	0	0	0	0	−0.2759598	1	2	1	0	0	−0.0005138
4	0	0	0	0	0.1781815	2	1	1	0	0	−0.0002668
0	2	0	0	0	0.4389737	2	2	0	0	0	−0.0067504
0	3	0	0	0	−0.4586550	2	0	2	0	0	−0.0034541
0	4	0	0	0	0.2953499	0	2	2	0	0	−0.0079609
0	0	2	0	0	0.1863816	3	0	1	0	0	−0.0004416
0	0	3	0	0	−0.1879897	3	1	0	0	0	−0.0045833
0	0	4	0	0	0.1266664	0	3	1	0	0	−0.0002983
0	0	0	2	0	0.0043264	1	3	0	0	0	0.0028970
0	0	0	4	0	0.0012367	0	1	3	0	0	−0.0018284
0	0	0	0	2	0.0374888	1	0	3	0	0	0.0011310
0	0	0	0	4	0.0122645	0	0	0	2	2	0.0011703
0	0	0	1	1	0.0014939	0	0	0	3	1	−0.0001832
1	1	0	0	0	0.0067488	0	0	0	1	3	−0.0019816
0	1	1	0	0	−0.0129058	2	0	0	2	0	−0.0054332
1	0	1	0	0	0.0048175	2	0	0	0	2	0.0019590
2	1	0	0	0	−0.0076502	0	2	0	2	0	0.0027339
1	2	0	0	0	−0.0101673	0	2	0	0	2	−0.0017913
2	0	1	0	0	−0.0025142	0	0	2	2	0	−0.0007879
1	0	2	0	0	−0.0002654	0	0	2	0	2	−0.0007593
0	2	1	0	0	0.0005100	2	0	0	1	1	−0.0110901
0	1	2	0	0	0.0022446	0	2	0	1	1	−0.0075187
1	1	1	0	0	−0.0002541	0	0	2	1	1	−0.0023881
1	0	0	2	0	0.0032940	1	1	0	1	1	0.0129357
1	0	0	0	2	−0.0070304	1	0	1	1	1	0.0017350
1	0	0	1	1	0.0163667	0	1	1	1	1	0.0026129
0	1	0	2	0	−0.0150710	1	1	0	2	0	−0.0146231
0	1	0	0	2	−0.0390815	1	1	0	0	2	−0.0042166
0	1	0	1	1	0.0045897	1	0	1	2	0	−0.0039763
0	0	1	2	0	−0.0007778	1	0	1	0	2	0.0029614
0	0	1	0	2	−0.0092509	0	1	1	2	0	0.0009467
0	0	1	1	1	−0.0004954	0	1	1	0	2	0.0078844

[a]Reference structure: $R_1^{ref} = 1.340$ Å, $R_2^{ref} = 1.236$ Å and $r^{ref} = 1.079$ Å

Table 18: Potential terms (in atomic units) for l-C$_3$H$^+$ as obtained by **Comp. 1**.[a]

							C_{ijklm}		
					CCSD(T*)	ΔCV	ΔDK	Δ(Q)-(T)	ΔQ-(Q)
i	j	k	l	m	/VQZ-F12	/CV6Z	/VQZ(-DK)	/VTZ	/VDZ
1	0	0	0	0	−0.0034637	0.0037423	0.0001910	−0.0018596	0.0005222
2	0	0	0	0	0.2760557	−0.0029168	−0.0003962	−0.0031538	0.0012244
3	0	0	0	0	−0.2805049	0.0014269	0.0002919	−0.0026042	0.0015128
4	0	0	0	0	0.1825856	−0.0006542	−0.0001310	−0.0017064	0.0011639
5	0	0	0	0	−0.0958850	0.0002102	0.0000485	−0.0010160	0.0010331
6	0	0	0	0	0.0397553	0.0001428	−0.0000149	−0.0003228	0.0008330
0	1	0	0	0	−0.0039001	0.0050644	0.0004361	−0.0016505	0.0006811
0	2	0	0	0	0.4462089	−0.0038761	−0.0006229	−0.0000521	0.0008485
0	3	0	0	0	−0.4657417	0.0021112	0.0003744	0.0000408	0.0004504
0	4	0	0	0	0.3001597	−0.0008592	−0.0001929	0.0005134	−0.0002229
0	5	0	0	0	−0.1604960	0.0003455	0.0000809	0.0002169	−0.0001057
0	6	0	0	0	0.0642770	−0.0001649	−0.0000240	−0.0004772	0.0002720
0	0	1	0	0	−0.0009923	0.0008822	0.0001004	0.0001407	0.0000124
0	0	2	0	0	0.1874114	−0.0007736	−0.0001365	−0.0000892	0.0000256
0	0	3	0	0	−0.1888689	0.0004794	0.0000908	−0.0000442	−0.0000122
0	0	4	0	0	0.1281412	−0.0002094	−0.0000465	−0.0000071	−0.0000037
0	0	5	0	0	−0.0735346	0.0000282	0.0000209	0.0000356	−0.0000018
0	0	6	0	0	0.0375901	0.0000010	−0.0000080	−0.0000412	0.0000276
0	0	7	0	0	−0.0226413				
0	0	8	0	0	0.0175528				
0	0	0	2	0	0.0040512	0.0001219	−0.0000335	0.0001495	0.0000644
0	0	0	4	0	−0.0001729	−0.0000089	0.0000000	0.0001935	−0.0000633
0	0	0	6	0	0.0000125	−0.0000014	0.0000021	−0.0000017	−0.0000236
0	0	0	8	0	0.0000835	−0.0000044	0.0000004	−0.0000141	0.0000153
0	0	0	0	2	0.0380040	−0.0000347	−0.0000266	−0.0003286	0.0001025
0	0	0	0	4	−0.0001527	0.0000008	0.0000007	−0.0002670	0.0000514
0	0	0	0	6	0.0004114	0.0000095	0.0000004	−0.0001489	0.0000508
1	1	0	0	0	0.0074475	0.0003428	−0.0000779	0.0002102	0.0000718
2	1	0	0	0	−0.0070165	−0.0001834	−0.0000309	−0.0015549	0.0002266
1	2	0	0	0	−0.0115880	−0.0001020	−0.0000222	0.0019297	0.0003842
1	0	1	0	0	0.0048856	0.0000229	0.0000002	−0.0000330	0.0000563
2	0	1	0	0	−0.0024191	−0.0000208	−0.0000035	−0.0000588	0.0000396
1	0	2	0	0	−0.0002482	0.0000071	−0.0000002	0.0000123	−0.0000295
0	1	1	0	0	−0.0132871	0.0001162	−0.0000328	0.0002886	−0.0000902
0	2	1	0	0	0.0007706	−0.0000111	−0.0000215	0.0000149	−0.0001714
0	1	2	0	0	0.0023561	−0.0000243	−0.0000108	−0.0000641	0.0000779
1	1	1	0	0	−0.0007780	0.0000038	−0.0000020	0.0002477	0.0000839
0	0	0	1	1	0.0011890	−0.0000168	0.0000274	−0.0000552	0.0001920
1	0	0	2	0	0.0032920	−0.0000951	−0.0000188	0.0004775	−0.0003602
1	0	0	0	2	−0.0070356	0.0000333	−0.0000160	−0.0000328	−0.0000200
1	0	0	1	1	0.0163469	−0.0001712	0.0000383	0.0001356	−0.0000422
0	1	0	2	0	−0.0147400	−0.0000479	−0.0000234	−0.0011424	0.0005198
0	1	0	0	2	−0.0387392	0.0001595	−0.0000418	−0.0003602	0.0003624
0	1	0	1	1	0.0041953	0.0001113	−0.0000171	−0.0002774	0.0005770
0	0	1	2	0	−0.0006428	−0.0000094	−0.0000027	−0.0000217	0.0000028
0	0	1	0	2	−0.0093026	0.0000295	−0.0000077	−0.0000434	−0.0000056
0	0	1	1	1	−0.0004604	0.0000150	−0.0000051	−0.0000767	0.0000195

[a]Reference structure: $R_1^{ref} = 1.340$ Å, $R_2^{ref} = 1.236$ Å and $r^{ref} = 1.079$ Å.
ΔDBOC/CVQZ: $C_{00100} = -0.0001470$ and $C_{00200} = -0.0001040$.

Table 19: Parameters (in atomic units) of the stretch-only EDMF for l-C_3H^+.

i	j	k	D_{ijk}	i	j	k	D_{ijk}
1	0	0	−1.48679	0	0	4	−0.03122
0	1	0	1.02864	1	1	0	0.06585
0	0	1	0.52690	2	1	0	0.23612
2	0	0	0.19976	1	2	0	−0.10613
3	0	0	0.08209	0	1	1	0.18477
4	0	0	0.03171	0	2	1	−0.06364
0	2	0	−0.35656	0	1	2	0.01259
0	3	0	0.00570	1	0	1	−0.03186
0	4	0	0.01206	2	0	1	0.02784
0	0	2	0.09896	1	0	2	−0.01911
0	0	3	0.00169	1	1	1	0.01387

A.3 Additional tables and force fields for C$_4$

Table 20: Calculated anharmonic constants χ_{ij} and $\chi_{l_t l_t}$ (in cm^{-1}) for C$_4$ as obtained by the "best estimate"+MRCI(-Q)-F12/VDZ-F12 and "best estimate"+ACPF-F12/VTZ-F12 force fields.[a]

		MRCI(+Q)-F12				ACPF-F12			
	$i\backslash j$	1	2	3	4	5	1	2	3	4	5
	1	-7.6					-9.8				
	2	-9.5	-1.3				-8.9	-1.5			
χ_{ij}	3	-21.0	-5.9	-5.6			-19.3	-6.4	-6.2		
	4	-15.8	18.8	1.2	-1.8		-19.4	16.8	-2.8	3.1	
	5	-5.4	4.7	-0.7	2.1	0.5	-5.7	4.6	-1.0	1.5	0.2
$\chi_{l_t l_t}$	4			3.3					1.7		
	5			0.7	0.4				0.7	0.5	

[a]The cubic and diagonal quartic force constants from **Comp. 1** and additional quartic force constants from the MRCI(+Q)-F12 or ACPF-F12 force-field were used along with the estimate ωs to derive these anharmonic constants.

Table 21: Non-redundant potential terms (in atomic units) up to cubic ones for C$_4$ as calculated by different multi-reference methods.[a]

					C_{ijklm} MRCI-F12 /VDZ-F12	ACPF-F12 /VTZ-F12	CIPT2(+Q) /CVTZ
i	j	k	l	m			
2	0	0	0	0	0.3054800	0.2903717	0.2905405
3	0	0	0	0	−0.3377661	−0.3246651	−0.3238890
0	2	0	0	0	0.3437433	0.3289466	0.3315502
0	3	0	0	0	−0.3599946	−0.3469693	−0.3495378
0	0	0	2	0	0.0198762	0.0193696	0.0169374
1	1	0	0	0	0.0308247	0.0274314	0.0279522
1	0	1	0	0	0.0135290	0.0144653	0.0144095
2	1	0	0	0	0.0044335	0.0022070	0.0018686
1	2	0	0	0	−0.0226171	−0.0248354	−0.0244623
2	0	1	0	0	−0.0092731	−0.0085123	−0.0082546
1	1	1	0	0	0.0103803	0.0101949	0.0102862
0	0	0	1	1	0.0018801	0.0018615	0.0028132
1	0	0	2	0	−0.0035387	−0.0001430	−0.0031612
1	0	0	0	2	−0.0024345	−0.0018065	−0.0009262
0	1	0	2	0	−0.0336998	−0.0313612	−0.0391378
1	0	0	1	1	0.0039304	0.0036032	0.0021941
0	1	0	1	1	0.0008835	0.0020336	0.0010401

[a]Reference structures:
MRCI-F12, ACPF-F12: $R_{\text{out}}^{\text{ref}} = 1.31131$ Å and $R_{\text{in}}^{\text{ref}} = 1.29108$ Å.
CIPT2(+Q): $R_{\text{out}}^{\text{ref}} = 1.3121$ Å and $R_{\text{in}}^{\text{ref}} = 1.2920$ Å.

Table 22: Non-redundant potential terms (in atomic units) for C_4 for the basis and smaller contributions of the **Comp. 1** method.[a]

						C_{ijklm}			
					UCCSD(T)	ΔCV	ΔDK	Δ(Q)-(T)	ΔQ-(Q)
i	j	k	l	m	/VQZ	CVTZ	/VTZ	/VTZ	/VDZ
1	0	0	0	0	−0.0009877	0.0029579	0.0002008	−0.0029871	
2	0	0	0	0	0.2905190	−0.0022229	−0.0004855	−0.0007557	
3	0	0	0	0	−0.3250822	0.0021472	0.0003341	0.0010430	
4	0	0	0	0	0.2043520	0.0028072		−0.0003337	
0	1	0	0	0	−0.0013294	0.0030805	0.0003109	−0.0023169	
0	2	0	0	0	0.3300984	−0.0022461	−0.0005221	−0.0019207	
0	3	0	0	0	−0.3463261	0.0002956	0.0002907	−0.0017013	
0	4	0	0	0	0.2179269	0.0008954		0.0002804	
0	5	0	0	0	−0.1185270				
0	0	0	2	0	0.0180114	0.0001320	−0.0000370	−0.0023856	0.0006303
0	0	0	4	0	0.0104800	−0.0001655	−0.0000221	0.0038502	
0	0	0	6	0	0.0046496	0.0001286	−0.0000100		
1	1	0	0	0	0.0277709	0.0002780	−0.0000662	0.0000859	
1	0	1	0	0	0.0170977	−0.0001000	0.0000242	−0.0038762	
2	1	0	0	0	0.0029960	−0.0021027	−0.0003353	−0.0045554	
1	2	0	0	0	−0.0252313	0.0007058	0.0005596	−0.0000585	
2	0	1	0	0	−0.0090016	0.0085384	0.0005596	0.0055866	
1	1	1	0	0	0.0120508	0.0007488	−0.0004848	0.0001175	
0	0	0	1	1	0.0019632	−0.0000807	0.0000137	0.0021202	
1	0	0	2	0	−0.0001115	0.0003002	0.0000068	0.0049340	
1	0	0	0	2	−0.0029961	0.0005133	0.0000068	0.0047947	
0	1	0	2	0	−0.0323161	−0.0004408	0.0000131	−0.0048554	
1	0	0	1	1	0.0025557	−0.0000984	−0.0000493	−0.0060001	
0	1	0	1	1	0.0006440	0.0001999	0.0000562	0.0044895	

[a]Reference structure: $R_{\text{out}}^{\text{ref}} = 1.3121$ Å and $R_{\text{in}}^{\text{ref}} = 1.2920$ Å.

Table 23: Non-redundant potential terms (in atomic units) up to quartic ones for C$_4$ as calculated by MRCI(+Q)-F12 and ACPF-F12.[a]

i	j	k	l	m	C_{ijklm} MRCI(+Q) /VDZ-F12	ACPF VTZ-F12	i	j	k	l	m	C_{ijklm} MRCI(+Q) /VDZ-F12	ACPF /VTZ-F12
2	0	0	0	0	0.2917434	0.2900535	1	3	0	0	0	−0.0283450	0.0033751
3	0	0	0	0	−0.3261305	−0.3241695	0	0	0	1	1	0.0023352	0.0023188
4	0	0	0	0	0.2113990	0.2018873	0	0	0	2	2	0.0032400	0.0111502
0	2	0	0	0	0.3320105	0.3285162	0	0	0	1	3	0.0036904	−0.0036290
0	3	0	0	0	−0.3488628	−0.3459026	1	0	0	2	0	−0.0015800	−0.0009034
0	4	0	0	0	0.2247906	0.2143714	1	0	0	0	2	−0.0009994	−0.0012309
0	0	0	2	0	0.0157732	0.0187757	0	1	0	2	0	−0.0349183	−0.0313323
0	0	0	4	0	0.0041348	0.0090225	0	1	0	1	1	0.0014250	0.0034958
1	1	0	0	0	0.0271886	0.0273800	0	1	0	1	3	0.0008347	0.0023396
1	0	1	0	0	0.0145121	0.0144803	2	0	0	2	0	−0.0269476	−0.0373090
2	1	0	0	0	0.0032572	0.0018744	0	2	0	2	0	0.0040001	−0.0100818
1	2	0	0	0	−0.0236519	−0.0248337	2	0	0	0	2	−0.0050013	−0.0243661
2	0	1	0	0	−0.0085269	−0.0086178	2	0	0	1	1	0.0036932	0.0235069
1	1	1	0	0	0.0097055	0.0094292	0	2	0	1	1	−0.0012805	0.0225245
2	2	0	0	0	0.0057481	−0.0387622	1	1	0	2	0	0.0148706	0.0015581
2	0	2	0	0	0.0105457	−0.0339371	1	1	0	0	2	0.0106055	0.0182539
2	1	1	0	0	0.0027551	−0.0051305	1	0	1	2	0	−0.0198983	−0.0163034
1	2	1	0	0	0.0209556	−0.0419237	1	1	0	1	1	−0.0147966	−0.0084280
3	1	0	0	0	0.0239323	−0.0100447	1	0	1	1	1	0.0232257	0.0188842
3	0	1	0	0	−0.0000933	0.0007984							

[a]Reference structure:
MRCI(+Q)-F12: $R_{out}^{ref} = 1.31131$ Å and $R_{in}^{ref} = 1.29108$ Å.
ACPF-F12: $R_{out}^{ref} = 1.3121$ Å and $R_{in}^{ref} = 1.2920$ Å.

Printed in the United States
By Bookmasters